endorsed for
Edexcel

Edexcel GCSE (9-1)
Mathematics
Foundation

Practice, Reasoning and Problem-solving Book

Confidence • Fluency • Problem-solving • Reasoning

Author: Katherine Pate

ALWAYS LEARNING

PEARSON

Published by Pearson Education Limited, 80 Strand, London WC2R 0RL.

www.pearsonschoolsandfecolleges.co.uk

Copies of official specifications for all Edexcel qualifications may be found on the website: www.edexcel.com

Text © Pearson Education Limited 2015
Edited by ProjectOne Publishing Solutions, Scotland
Typeset and illustrated by Tech-set Ltd, Gateshead
Original illustrations © Pearson Education Limited 2015

The right of Katherine Pate be identified as the author of this work has been asserted by her in accordance with the Copyright, Designs and Patents Act 1988.

First published 2015

18 17 16 15
10 9 8 7 6 5 4 3 2 1

British Library Cataloguing in Publication Data
A catalogue record for this book is available from the British Library

ISBN 978 1 292 10017 3

Printed in Slovakia by Neografia

Acknowledgements
The publisher would like to thank the following for their kind permission to reproduce their photographs:

Cover images: Front: Created by **Fusako**, Photography by NanaAkua

Every effort has been made to contact copyright holders of material reproduced in this book. Any omissions will be rectified in subsequent printings if notice is given to the publishers.

A note from the publisher
In order to ensure that this resource offers high-quality support for the associated Pearson qualification, it has been through a review process by the awarding body. This process confirms that this resource fully covers the teaching and learning content of the specification or part of a specification at which it is aimed. It also confirms that it demonstrates an appropriate balance between the development of subject skills, knowledge and understanding, in addition to preparation for assessment.

Endorsement does not cover any guidance on assessment activities or processes (e.g. practice questions or advice on how to answer assessment questions), included in the resource nor does it prescribe any particular approach to the teaching or delivery of a related course.

While the publishers have made every attempt to ensure that advice on the qualification and its assessment is accurate, the official specification and associated assessment guidance materials are the only authoritative source of information and should always be referred to for definitive guidance.

Pearson examiners have not contributed to any sections in this resource relevant to examination papers for which they have responsibility.

Examiners will not use endorsed resources as a source of material for any assessment set by Pearson.

Endorsement of a resource does not mean that the resource is required to achieve this Pearson qualification, nor does it mean that it is the only suitable material available to support the qualification, and any resource lists produced by the awarding body shall include this and other appropriate resources.

Contents

Welcome to Edexcel GCSE (9-1) Mathematics Foundation Booster Practice, Reasoning and Problem-solving Book

This Booster Book is packed with extra practice on all the new and most demanding content of the new GCSE Specification for Foundation tier giving you more opportunities to practise answering questions to gain confidence and develop problem-solving and reasoning skills.

There are sections relating to each strand of the specification, carefully ordered to optimise the reuse of key skills from the Number and Algebra strands.

References to the *Pearson Edexcel GCSE (9-1) Foundation Student Book* show when you can attempt a section and where to go for extra support.

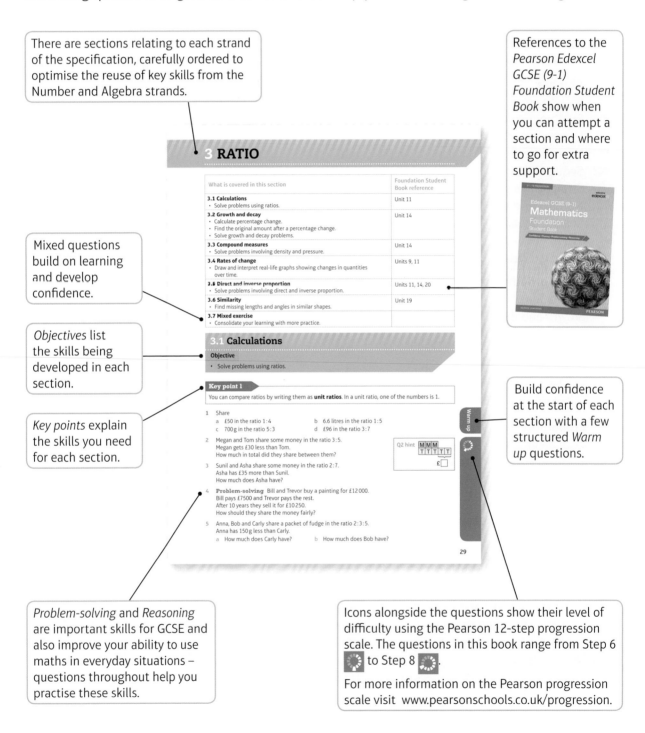

Mixed questions build on learning and develop confidence.

Objectives list the skills being developed in each section.

Key points explain the skills you need for each section.

Build confidence at the start of each section with a few structured *Warm up* questions.

3 RATIO

What is covered in this section	Foundation Student Book reference
3.1 Calculations • Solve problems using ratios.	Unit 11
3.2 Growth and decay • Calculate percentage change. • Find the original amount after a percentage change. • Solve growth and decay problems.	Unit 14
3.3 Compound measures • Solve problems involving density and pressure.	Unit 14
3.4 Rates of change • Draw and interpret real-life graphs showing changes in quantities over time.	Units 9, 11
3.5 Direct and inverse proportion • Solve problems involving direct and inverse proportion.	Units 11, 14, 20
3.6 Similarity • Find missing lengths and angles in similar shapes.	Unit 19
3.7 Mixed exercise • Consolidate your learning with more practice.	

3.1 Calculations

Objective

• Solve problems using ratios.

Key point 1

You can compare ratios by writing them as **unit ratios**. In a unit ratio, one of the numbers is 1.

1 Share
  a £50 in the ratio 1:4   b 6.6 litres in the ratio 1:5
  c 700g in the ratio 5:3   d £96 in the ratio 3:7

2 Megan and Tom share some money in the ratio 3:5.
  Megan gets £30 less than Tom.
  How much in total did they share between them?

Q2 hint
M M M
T T T T T
£ ☐

3 Sunil and Asha share some money in the ratio 2:7.
  Asha has £35 more than Sunil.
  How much does Asha have?

4 **Problem-solving** Bill and Trevor buy a painting for £12 000.
  Bill pays £7500 and Trevor pays the rest.
  After 10 years they sell it for £10 250.
  How should they share the money fairly?

5 Anna, Bob and Carly share a packet of fudge in the ratio 2:3:5.
  Anna has 150g less than Carly.
  a How much does Carly have?   b How much does Bob have?

29

Problem-solving and *Reasoning* are important skills for GCSE and also improve your ability to use maths in everyday situations – questions throughout help you practise these skills.

Icons alongside the questions show their level of difficulty using the Pearson 12-step progression scale. The questions in this book range from Step 6 to Step 8.

For more information on the Pearson progression scale visit www.pearsonschools.co.uk/progression.

1 NUMBER

What is covered in this section	Foundation Student Book reference
1.1 Calculations • Estimate answers to calculations. • Write an inequality to represent an error interval. • Use a calculator for complex calculations.	Units 1, 17
1.2 Factors and multiples • Write a number as a product of its prime factors. • Use prime factor decomposition to find the HCF and LCM of two numbers.	Unit 1
1.3 Fractions • Add, subtract, multiply and divide mixed numbers.	Units 4, 18
1.4 Indices, powers and roots • Simplify expressions involving indices, roots and surds. • Understand and use zero and negative indices.	Units 1, 18
1.5 Standard form • Write very small and very large numbers in standard form. • Calculate with numbers in standard form.	Unit 18
1.6 Mixed exercise • Consolidate your learning with more practice.	

1.1 Calculations

Objectives

• Estimate answers to calculations.
• Write an inequality to represent an error interval.
• Use a calculator for complex calculations.

Key point 1

Measurements given to the nearest whole unit may be inaccurate by up to one half of a unit below and one half of a unit above. For example, the range of possible values for a length given as 3 cm to the nearest cm is 2.5 cm ≤ length < 3.5 cm.

Key point 2

To estimate the answer to a calculation, you can round every number to 1 significant figure (s.f.).

1   Round these numbers to 1 s.f.

    a   456   b   24   c   9.1   d   7.58

2   Estimate the answer to these calculations by first rounding each value to 1 s.f.

    a $\dfrac{1.7 + 12.2}{5.9}$ b $\sqrt{84.37 + 18.6}$

3   Write an inequality to show the possible values for

    a   680 (rounded to the nearest 10)

    b   1.27 (rounded to 2 d.p.)

    c   17.2 (rounded to 3 s.f.)

    d   200 (rounded to 2 s.f.)

Q3 hint $\square \leqslant n < \square$

4 A machine produces rolls of paper. Each roll is 20 m to the nearest centimetre.

    a What is the minimum possible length of paper in the roll?

    b Write an inequality to show the possible lengths of paper, l, in a roll.

5 Shampoo bottles contain 240 ml of shampoo. There is an error of ±5% in the volume of shampoo in the bottle. Work out the minimum and maximum possible volumes of shampoo in the bottle.

6 **Problem-solving** A machine fills packs of crisps with 120 g of crisps. There is an error of ±3% in the mass of the crisps in the packs. The crisps label says, 'Minimum contents 115 g'. Is this true? Show how you get your answer.

> **Q6** A common error is to write 'Yes' or 'No' and to not show calculations to explain why.

7 Tom calculates that $892.17 \div 18.4 = 85.8$ (1 d.p.). Estimate the answer to his calculation to show he is wrong.

8 Use estimation to see which of these calculations is wrong and which could be correct.

    a $\dfrac{3.2^2 + 9.17}{6.19 - 2.3} = 4.99$ (3 s.f.)   **b** $\sqrt{\dfrac{2.17^2 + 57.2}{3.9^2}} = 3.98$ (3 s.f.)

    c $\dfrac{5436 - 82^2}{5.36 + 12.97} = 149$ (3 s.f.)   **d** $\dfrac{9.8^3 \times 3.7}{1.8 \times 5.2^2} = 71.5$ (3 s.f.)

9 Use a calculator to work out the correct answer to the calculations in **Q8** that were wrong. Round to 3 s.f.

> **Q9** A common error is to round to 3 d.p. instead of 3 s.f.

10 Use a calculator to work these out. Round your answers to the number of s.f. given.

    a $\sqrt{\dfrac{4.1^2 + 2.15^3}{3.8^2}}$ (2 s.f.)   **b** $\sqrt[3]{\dfrac{24 - 3.29^2}{11.7 + 3.2}}$ (3 s.f.)

    c $\sqrt[3]{\dfrac{5^2 + 13.2^3}{19.2}}$ (3 s.f.)   **d** $\dfrac{3 + \sqrt{(-3)^2 + 2 \times 4 \times 3}}{6}$ (4 s.f.)

> **Q10 hint** Estimate the answers first, in order to check your calculations.

1.2 Factors and multiples

Objectives

* Write a number as a product of its prime factors.
* Use prime factor decomposition to find the HCF and LCM of two numbers.

Key point 3

All numbers can be written as a product of prime factors. This is called **prime factor decomposition**.

1 Write as a product of powers
    $2 \times 5 \times 3 \times 5 \times 2 \times 2$

> **Q1 hint** $2^\square \times 3^\square \times 5^\square$

2 Find

    a the HCF of 12 and 32

    b the LCM of 6 and 10.

3 **a** Copy and complete these factor trees.

    b Write 84 and 105 as products of their prime factors.

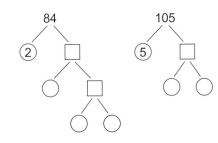

> **Q3 hint** Write the products of factors using index notation, smallest factor first.

4   What is the prime factor decomposition of 72?

5   Copy and complete this Venn diagram to find the HCF and LCM of 84 and 72.

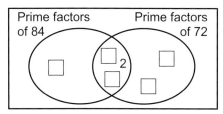

Q5 A common error is to give the LCM answer for the HCF or the HCF for the LCM.

6   Find the HCF and LCM of
    a   105 and 84   b   28 and 70   c   132 and 60

7   Two numbers have HCF 8 and LCM 120.
    What are the two numbers?

Q7 hint

$\square \times 8 \times \square = 120$

8   **Problem-solving** Find two numbers with LCM 70 and HCF 5.

9   $A = 2^2 \times 5 \times 7$ $B = 3 \times 5^2 \times 11$
    $C = 2^2 \times 13$ $D = 2 \times 3^2 \times 7^2$
    a   Find the HCF of
        i A and D ii A and C iii B and D
    b   Which two numbers have LCM
        i   1820   ii   42 900   iii   8820
    c   Which two numbers have no common factors?

Q9 hint Draw Venn diagrams.

10 **Problem-solving** Number 6 buses stop at the railway station every 9 minutes.
    Number 15 buses stop at the railway station every 12 minutes.
    At 10 am a number 6 bus and a number 15 bus both stop at the station.
    When will a number 6 bus and a number 15 bus next stop at the station together?

Q10 A common error is to not realise that this is an LCM question.

1.3 Fractions

Objective

• Add, subtract, multiply and divide mixed numbers.

Key point 4

To add, subtract, multiply or divide mixed numbers, you can write them as improper fractions first.

1   Work out
    a $\frac{5}{6} + \frac{3}{5}$ b $\frac{4}{9} - \frac{1}{6}$ c $\frac{3}{4} \times \frac{5}{7}$ d $\frac{2}{3} \div \frac{3}{8}$
    Give your answers as mixed numbers where appropriate.

2   Cancel then multiply
    a $\frac{5}{6} \times \frac{3}{7}$ b $\frac{4}{7} \times \frac{3}{8}$
    c $\frac{6}{12} \times \frac{2}{3}$ d $\frac{7}{9} \times \frac{3}{14}$

Q2 hint $\dfrac{5 \times 3}{6 \times 7} = \dfrac{{}^{1}\cancel{3} \times 5}{{}^{2}\cancel{6} \times 7} = \dfrac{\square}{\square}$

3   Work out
    a $2\frac{1}{5} + 1\frac{2}{5}$ b $3\frac{1}{7} + 2\frac{3}{5}$
    c $5\frac{1}{4} + 1\frac{4}{9}$ d $4\frac{3}{10} + 1\frac{1}{3}$

Q3 hint Write your answers as mixed numbers.

Warm up

4 Work out

a $2\frac{3}{5} + 1\frac{7}{10}$ b $3\frac{5}{8} + 4\frac{3}{4}$

c $5\frac{6}{7} + 2\frac{2}{3}$ d $1\frac{11}{12} + 3\frac{5}{9}$

> **Q4** A common error when adding or subtracting fractions is to add/subtract the numerators and denominators.

5 Work out

a $5\frac{4}{9} - 1\frac{1}{5}$ b $3\frac{3}{4} - 1\frac{5}{8}$ c $6\frac{6}{7} - 3\frac{1}{3}$ d $4\frac{8}{11} - 3\frac{2}{5}$

6 Work out

a $4\frac{1}{4} - 1\frac{3}{5}$ b $3\frac{1}{7} - 2\frac{1}{4}$ c $5\frac{2}{3} - 2\frac{7}{9}$ d $4\frac{5}{12} - 2\frac{2}{3}$

7 Work out

a $3\frac{1}{5} \times \frac{2}{3}$ b $\frac{2}{5} \times 2\frac{1}{4}$

c $\frac{5}{7} \times 3\frac{1}{2}$ d $\frac{3}{4} \times 2\frac{2}{5}$

> **Q7** A common error is to not change to improper fractions first.

8 Work out

a $2\frac{1}{4} \div \frac{1}{5}$ b $1\frac{1}{2} \div \frac{4}{7}$

c $3\frac{2}{3} \div \frac{5}{6}$ d $1\frac{7}{8} \div \frac{5}{8}$

> **Q8** A common error when dividing fractions is to forget to 'flip' the fraction you are dividing by, when you change the calculation to multiplication.

9 Work out

a $2\frac{1}{2} \times 1\frac{1}{6}$ b $1\frac{2}{3} \times 1\frac{1}{8}$ c $1\frac{1}{7} \times 3\frac{3}{4}$ d $3\frac{1}{5} \times 2\frac{1}{2}$

10 Work out

a $2\frac{3}{4} : 2\frac{1}{3}$ b $1\frac{4}{5} \div 1\frac{3}{4}$ c $2\frac{1}{4} \div 3\frac{3}{5}$ d $2\frac{2}{3} \div 1\frac{7}{9}$

11 **Problem-solving** Karl has these bags of sand.
He needs 8 kg of sand to make cement.
Does he have enough?
Show working to explain.

 $1\frac{5}{8}$ kg $2\frac{11}{12}$ kg $3\frac{9}{10}$ kg

12 **Problem-solving** Ailsa has $8\frac{3}{5}$ metres of wire.
a   How many $1\frac{1}{2}$ metre lengths can she cut from it?
b   How much is left over?

1.4 Indices, powers and roots

Objectives

*   Simplify expressions involving indices, roots and surds.
*   Understand and use zero and negative indices.

Key point 5

Any number (or term) raised to the power 0 is equal to 1.
Any number (or term) raised to the power −1 is the reciprocal of the number.
To raise a power of a number (or term) to another power, multiply the indices.

Key point 6

3^{-2} is the same as $(3^{-1})^2$.
To find a negative power, find the reciprocal and then raise the value to the positive power.

Key point 7

Expressions with square roots like $3\sqrt{2}$ are in **surd** form. $3\sqrt{2}$ means $3 \times \sqrt{2}$.
An answer in surd form is exact.

1 Work out

   a $3^2 \times 3^5$
   b 4^5
   c $\dfrac{5^6 \times 5^3}{5^5}$
   d $\dfrac{x^3 \times x}{x^2}$

   e $\left(\dfrac{1}{4}\right)^2$
   f $\left(\dfrac{2}{3}\right)^3$
   g $\left(\dfrac{1}{y}\right)^5$
   h $\left(\dfrac{x}{5}\right)^2$

2 Write the reciprocal of

   a $\dfrac{1}{4}$
   b $\dfrac{1}{3}$
   c 2
   d $\dfrac{3}{2}$

3 Write as a single power

   a $(5^2)^3$
   b $(2^3)^4$
   c $(6^3)^3$

   d $(10^5)^2$
   e $(x^4)^2$
   f $(t^4)^3$

   g $(m^2)^4$
   h $(n^3)^2$

> **Q3a hint** $(5^2)^3 = 5^{\square \times \square}$

> **Q3** A common error is to add the indices instead of multiplying.

4 Write as a single power

   a $(2^3 \times 2^4)^2$
   b $\left(\dfrac{4^3}{4}\right)^5$
   c $(5^3)^2 \times 5$
   d $\dfrac{10^5}{10^2} \times 10^3$

   e $\dfrac{3^5 \times 3^4}{(3^4)^2}$
   f $(2^3)^2 \times (2^2)^3$
   g $(x^3)^5 \times x$
   h $\dfrac{n^5 \times n^3}{(n^2)^2}$

5 Write without powers

   a 3^{-1}
   b 2^{-1}
   c 10^{-1}
   d y^{-1}

   e $\left(\dfrac{1}{4}\right)^{-1}$
   f $\left(\dfrac{2}{5}\right)^{-1}$
   g $\left(\dfrac{1}{z}\right)^{-1}$
   h $\left(\dfrac{a}{b}\right)^{-1}$

6 Evaluate

   a 2^0
   b 10^0
   c $\dfrac{2^4}{2^5}$

   d $\dfrac{10^2 \times 10^3}{10^5}$
   e $\dfrac{5^2 \times 5^3}{5^6}$
   f $3^4 \times 3^{-2}$

   g $6^{-1} \times 6^3$
   h $10^{-5} \times 10^4$

> **Q6 communication hint**
> **Evaluate** means 'work out the value of'.

7 Simplify

   a $\dfrac{x \times x^3}{x^4}$
   b $\dfrac{t^3}{t^4}$
   c $\dfrac{n^2 \times n^3}{n^6}$
   d $m^5 \times m^{-4}$

8 Write these in order, smallest first.

   8^0 $\left(\dfrac{1}{4}\right)^{-1}$ 6^{-1} $\left(\dfrac{2}{5}\right)^{-1}$ 2^{-1}

> **Q8** A common error is to not use the original expressions in the final ordering.

9 **Reasoning** Which of these are surds?

   a $\sqrt{36}$
   b $\sqrt{7}$
   c $\sqrt{11}$

   d $\sqrt{64}$
   e $\sqrt{46}$
   f $\sqrt{55}$

   g $\sqrt{81}$
   h $\sqrt{19}$

> **Q9 hint** $\sqrt{36}$ is not a surd because $\sqrt{36} = 6$.

10 Work out

   a $\sqrt{2} \times \sqrt{2}$
   b $(\sqrt{5})^2$
   c $(\sqrt{7})^2$

   d $(\sqrt{n})^2$
   e $4\sqrt{3} \times \sqrt{3}$
   f $\sqrt{5} \times 3\sqrt{5}$

   g $4\sqrt{2} \times 3\sqrt{2}$

> **Q10e hint**
> $4\sqrt{3} \times \sqrt{3} = 4 \times \sqrt{3} \times \sqrt{3}$

11 Simplify

   a $2\sqrt{3} + 5\sqrt{3}$
   b $3\sqrt{5} - 2\sqrt{5}$
   c $5\sqrt{17} + \sqrt{17}$
   d $\sqrt{11} + \sqrt{11}$

> **Q11a hint** $2\square + 5\square = 7\square$

12 Evaluate

 a 2^{-4} b 3^{-5} c 5^{-3}

 d 7^{-2} e 10^{-2} f $\left(\dfrac{1}{3}\right)^{-2}$ g $\left(\dfrac{1}{4}\right)^{-2}$

 h $\left(\dfrac{1}{3}\right)^{-3}$ i $\left(\dfrac{3}{5}\right)^{-2}$ j $\left(\dfrac{2}{7}\right)^{-2}$ k $\left(\dfrac{5}{2}\right)^{-3}$

> **Q12a hint** $2^{-4} = \dfrac{1}{2^{\square}} = \dfrac{1}{\square}$

13 Write as a single power

 a $\left(\dfrac{1}{n}\right)^{-2}$ b $\left(\dfrac{c}{d}\right)^{-3}$ c $\dfrac{x^2 \times x}{x^5}$

 d $r^{-6} \times r^4$ e $t^3 \div t^{-2}$ f $x^{-1} \times x^{-3}$

 g $y^3 \times \left(\dfrac{1}{y}\right)^2$ h $\left(\dfrac{1}{z}\right)^3 \times z$

> **Q13 hint** Write answers with negative powers as $\dfrac{1}{x^{\square}}$

14 Solve these equations.

 a $2^m = 32$ b $10^c = 10\,000$ c $3^d = \dfrac{1}{9}$ d $5 \times 2^x = 20$

 e $12 \times 2^y = 6$ f $4 \times 4^n = 64$ g $\dfrac{10^z}{2} = 0.005$

1.5 Standard form

Objectives

- Write very small and very large numbers in standard form.
- Calculate with numbers in standard form.

Key point 8

Standard form is a way of writing very large or very small numbers.
A number in standard form looks like this:

$$8.4 \times 10^5$$

This part is a number between 1 and 10   This part is a power of 10

Key point 9

To write a large number in standard form:
- place the decimal point after the first digit
- multiply by the power of 10 needed to give the original number.

Key point 10

To write a small number in standard form:
- place the decimal point after the first non-zero digit
- count how many places this has moved the digit to the left – this is the negative power of 10.

Key point 11

To multiply or divide numbers in standard form, multiply or divide the number parts as usual and use the laws of indices to simplify the power of 10.

Key point 12

To add or subtract numbers in standard form, write both numbers as ordinary numbers, add or subtract, and then convert back to standard form.

1. Write as a single power.
   - a $10^5 \times 10^3$
   - b 10×10^4
   - c $10^{-2} \times 10^4$
   - d $10^{-1} \times 10^{-5}$

2. Write as a decimal or ordinary number.
   - a 10^{-3}
   - b 10^{-2}
   - c 10^5
   - d 10^{-1}

3. Write as a power of 10.
   - a 0.0001
   - b 1000
   - c 0.00001
   - d 1 000 000
   - e 0.000 001
   - f 10 000

4. Write as a decimal or ordinary number.
   - a 7×10^5
   - b 1.6×10^5
   - c 2.37×10^4
   - d 3.6205×10^2
   - e $5.290\,26 \times 10^6$
   - f 4.1903×10^5
   - g 9.76×10^2
   - h 1.26×10^5

5. Write each number in standard form.

   > **Q5a hint** $2000 = 2 \times 1000 = 2 \times 10^{\square}$

   - a 2000
   - b 800 000
   - c 5600
   - d 92 650
   - e 125 400
   - f 125.3
   - g 21 591 000
   - h 5.4 million

   > **Q5** A common error is to write the first part either as a number less than 1 or as a number greater than 10.

6. Write each number in standard form.
   - a 0.0052
   - b 0.000 004 31
   - c 0.006 59
   - d 0.000 471
   - e 0.005 080 2
   - f 0.000 095
   - g 0.000 000 14
   - h 0.0237

7. Write each number as an ordinary number.
   - a 2.7×10^{-5}
   - b 3.46×10^{-3}
   - c 9.201×10^{-2}
   - d 8.4×10^{-4}
   - e 5.06×10^{-10}
   - f 1.72×10^{-6}
   - g 6.54×10^{-8}
   - h 5.192×10^{-7}

   > **Q7** A common error is to write the wrong number of zeros. Count them carefully.

8. **Reasoning** Is each of these in standard form? If not, rewrite in standard form.
   - a 1.3×10^4
   - b 32.6×10^{-2}
   - c 176.3×10^{-3}
   - d 10.5×10^3

   > **Q8b hint** $3.26 \times 10 \times 10^{-2}$

9. Giving your answers in standard form, work out
   - a $3 \times 10^4 \times 2 \times 10^3$
   - b $1.5 \times 10^{-2} \times 4 \times 10^6$
   - c $2 \times 10^5 \times 1.6 \times 10^{-3}$
   - d $(2 \times 10^4)^2$
   - e $5 \times 10^3 \times 7 \times 10^2$
   - f $2.5 \times 10^{-6} \times 6 \times 10^7$
   - g $(6 \times 10^{-3})^2$
   - h $7 \times 10^5 \times 3 \times 10^{-8}$

10. Giving your answers in standard form, work out
    - a $\dfrac{9 \times 10^5}{3 \times 10^2}$
    - b $\dfrac{7 \times 10^{-3}}{2 \times 10^2}$
    - c $\dfrac{6 \times 10^2}{3 \times 10^{-4}}$
    - d $\dfrac{5.4 \times 10^{-3}}{2.7 \times 10^{-2}}$
    - e $\dfrac{8.5 \times 10^6}{3.2 \times 10^2}$

    > **Q10e hint** Use a calculator for $8.5 \div 3.2$

    - f $(5.6 \times 10^{-3}) \div (1.75 \times 10^{-5})$
    - g $(8.72 \times 10^4) \div (4.5 \times 10^2)$

    > **Q10g hint** Give the decimal part to 2 decimal places.

    - h $(7.31 \times 10^{-4}) \div (1.5 \times 10^{-6})$

11. Work out
    - a $(4 \times 10^3) + (3 \times 10^4)$
    - b $(7 \times 10^{-2}) + (5 \times 10^{-3})$
    - c $(6.1 \times 10^4) + (4 \times 10^3)$
    - d $(9.5 \times 10^{-3}) + (2 \times 10^{-2})$
    - e $(8.256 \times 10^3) + (1.3 \times 10^{-2})$
    - f $(5.04 \times 10^6) + (9.1 \times 10^4)$
    - g $(3.452 \times 10^{-2}) + (1.07 \times 10^{-3})$
    - h $(2.75 \times 10^7) + (3.1 \times 10^3)$

12. Work out
    - a $(2 \times 10^9) - (3 \times 10^8)$
    - b $(5 \times 10^{-2}) - (6 \times 10^{-3})$
    - c $(4.1 \times 10^2) - (5.6 \times 10^{-1})$
    - d $(7.1 \times 10^8) - (3.5 \times 10^6)$
    - e $(4.27 \times 10^{-2}) - (8.1 \times 10^{-4})$
    - f $(8.46 \times 10^8) - (9.2 \times 10^2)$
    - g $(5.27 \times 10^{-3}) - (1.2 \times 10^{-5})$
    - h $(9.76 \times 10^{15}) - (8.1 \times 10^{12})$

1.6 Mixed exercise

Objective

• Consolidate your learning with more practice.

1 **Exam question**

Buses to Acton leave a bus station every 24 minutes.

Buses to Barton leave the same bus station every 20 minutes.

A bus to Acton and a bus to Barton both leave the bus station at 9 00 am.

When will a bus to Acton and a bus to Barton next leave the bus station at the same time? **(3 marks)**

June 2012, Q7, 1MA0/1H

Exam hint

A lot of students made mistakes adding times, or converting minutes to hours and minutes in this question.

2 **Exam question**

Work out an estimate for $\dfrac{31 \times 9.87}{0.509}$ **(3 marks)**

November 2012, Q5, 1MA0/1H

Exam hint

Working out the calculation exactly and then rounding gets no marks.

3 **Exam question**

Margaret has some goats.

The goats produce an average total of 21.7 litres of milk per day for 280 days.

Margaret sells the milk in $\frac{1}{2}$ litre bottles.

Work out an estimate for the total number of bottles that Margaret will be able to fill with the milk.

You must show clearly how you got your estimate. **(3 marks)**

June 2013, Q8, 1MA0/1H

Exam hint

Common errors in answering this question were not rounding and making mistakes with division calculations.

4 **Reasoning** a Write 42 and 504 as products of prime factors.

b Is 42 a factor of 504?

Explain how you know.

5 **Exam question**

Rita is going to make some cheeseburgers for a party.

She buys some packets of cheese slices and some boxes of burgers.

There are 20 cheese slices in each packet.

There are 12 burgers in each box.

Rita buys exactly the same number of cheese slices and burgers.

i How many packets of cheese slices and how many boxes of burgers does she buy?

Rita wants to put one cheese slice and one burger into each bread roll.

She wants to use all the cheese slices and all the burgers.

ii How many bread rolls does Rita need? **(4 marks)**

November 2013, Q7, 1MA0/1H

Exam hint

Some candidates wrote out lists of numbers for this question, and made mistakes in their lists.

6 **Reasoning** Ravi is joining two pieces of metal
with a bolt, as shown in the diagram.
He has three bolts to choose from.

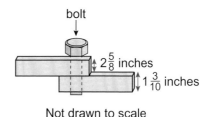

bolt

$2\frac{5}{8}$ inches

$1\frac{3}{10}$ inches

Not drawn to scale

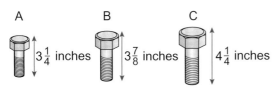

A $3\frac{1}{4}$ inches   B $3\frac{7}{8}$ inches   C $4\frac{1}{4}$ inches

Which bolt should he use?
Show working to explain.

Q7 communication hint
Hence means use your answer
from the first part to help you.

7 **Reasoning** a Find the HCF of 770, 330 and 1155.
   b Hence write the ratio 770 : 330 : 1155 in its simplest form.

8 Calculate the area of this rectangle in square yards.

$2\frac{3}{4}$ yards

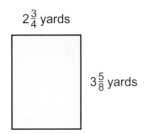

$3\frac{5}{8}$ yards

9 **Reasoning** The number 441 written as a product of its prime factors is
$441 = 3^2 \times 7^2$
   a Write $\sqrt{441}$ as a product of its prime factors.
   b Write $\sqrt{225}$ as a product of its prime factors.
   c Write $\sqrt[3]{2744}$ as a product of its prime factors.

10 **Problem-solving** This rectangle has area $8\frac{7}{12}$ square feet.

Area $= 8\frac{7}{12}$ square feet $2\frac{1}{6}$ feet

Calculate its length, in feet.

11 **Exam question**

One sheet of paper is 9×10^{-3} cm thick.
Mark wants to put 500 sheets of paper into the paper tray of
his printer.
The paper tray is 4 cm deep.
Is the paper tray deep enough for 500 sheets of paper?
You must explain your answer.   **(3 marks)**

June 2013, Q15, 1MA0/1H

Exam hint
You need to show calculations
that explain why you decided
on 'Yes' or 'No' for your
answer.

12 **Problem-solving** In the 2001 Census the population of London was 8.294×10^6 and the
population of Bristol was 4.28×10^5.
Approximately how many times bigger was the population of London than the population
of Bristol?

13 Problem-solving How many 5 MB song files can you fit on a 2 GB memory stick?

Q13 hint 1 MB = 10^6 bytes, 1 GB = 10^9 bytes.

14 Give your answer to each calculation in standard form.
   a $(2 \times 10^2)^3$
   b $(4 \times 10^5)^3$
   c $(3 \times 10^{-3})^3$
   d $(5 \times 10^{-1})^3$

15 Problem-solving Which of these numbers are cubes of whole numbers?

Q15 hint Write as a whole cube number × $10^\square$

   a 8×10^5
   b 2.7×10^7
   c 6.4×10^{-5}
   d 1×10^7
   e 1.25×10^8
   f 6.4×10^4

16 Calculate the area of this triangle, giving your answer in standard form.

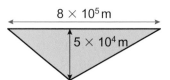

8×10^5 m

5×10^4 m

17 ┌─ **Exam question** ─┐

Competition

a prize every 2014 seconds

In a competition, a prize is won every 2014 seconds.
Work out an estimate for the number of prizes won in 24 hours.
You must show your working. **(4 marks)**

June 2014, Q13, 1MA0/1H

Exam hint
Exact calculations get no marks – the question asks you to estimate.

18 ┌─ **Exam question** ─┐

Write the following numbers in order of size.
Start with the smallest number.

0.038×10^2 3800×10^{-4} 380 0.38×10^{-1}

(2 marks)

November 2012, Q20, 1MA0/1H

Exam hint
Think about the powers of 10, or you could convert to ordinary numbers. The most common mistake students made was putting 3800×10^{-4} in the wrong place.

19 ┌─ **Exam question** ─┐

   a Write down the reciprocal of 5. **(1 mark)**
   b Evaluate 3^{-2} **(1 mark)**
   c Calculate $9 \times 10^4 \times 3 \times 10^3$
      Give your answer in standard form. **(2 marks)**

November 2013, Q14, 1MA0/1H

Exam hint
Many candidates did not know the meaning of 'reciprocal'.

20 Giving your answers as mixed numbers where appropriate, work out
   a $(1\frac{3}{7})^2$
   b $(2\frac{4}{5})^2$
   c $(1\frac{1}{3})^3$
   d $(1\frac{1}{4})^3$

21 The distance to a star is estimated as 4.0×10^{18} km. There is an error of ±2% in the distance to the star. Find the smallest and largest possible distances.

22 Simplify
   a $\dfrac{6\sqrt{2}}{3} + 5\sqrt{2}$
   b $\sqrt{8^2 - 5^2}$
   c $\sqrt{3^2 + 1^2}$
   d $\dfrac{3\sqrt{5} + 7\sqrt{5}}{5}$

2 ALGEBRA

What is covered in this section	Foundation Student Book reference
2.1 Writing equations and formulae • Write equations and formulae. • Solve equations.	Unit 5
2.2 Rearranging formulae • Change the subject of a formula.	Units 5, 20
2.3 Expanding and factorising • Multiply double brackets. • Factorise quadratic expressions.	Unit 16
2.4 Inequalities • Solve linear inequalities. • Show solutions to inequalities on a number line.	Unit 5
2.5 Sequences • Recognise and use Fibonacci-type sequences, quadratic sequences and geometric sequences. • Find and use the nth term of an arithmetic sequence.	Unit 5
2.6 Straight-line graphs • Find the equation of a straight-line graph, and of a line joining two points. • Interpret the gradient and y-intercept of a graph in context.	Units 9, 20
2.7 Non-linear graphs • Draw quadratic, cubic and reciprocal graphs. • Estimate solutions to an equation from its graph.	Units 16, 20
2.8 Solving equations • Solve quadratic equations algebraically. • Solve simultaneous equations using graphs or algebra.	Units 16, 20
2.9 Proof • Recognise equations and identities. • Prove results using algebra.	Unit 20
2.10 Mixed exercise • Consolidate your learning with more practice.	

2.1 Writing equations and formulae

Objectives

- Write equations and formulae.
- Solve equations.

Key point 1

An **equation** contains an unknown number (a letter) and an '=' sign.
When you solve an equation you work out the value of the unknown number.

Key point 2

A **formula** shows the relationship between two or more variables (letters).
You can use substitution to find an unknown value.

1   Write a formula for the perimeter, P, of each triangle.

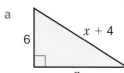   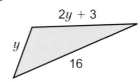

a   b $3a - 5$ c $2y + 3$

6        $x + 4$ a y

x 16

Q1a hint
Start $P = \ldots$
Simplify the
right-hand side.

2   Write an equation using n as the unknown starting number.
    Solve your equation to find n.
    a   I think of a number, multiply it by 7 and add 4. My answer is 25.
    b   I think of a number, divide it by 3 and subtract 2. My answer is 2.
    c   I think of a number, subtract 4 and then multiply by 5. My answer is 30.
    d   I think of a number, add 8 and then divide by 2. My answer is 7.

3   Anna is paid £7 per hour for babysitting before 11 pm and £10 per hour after 11 pm.
    a   Write a formula for her pay, P, for a hours before 11 pm and b hours after 11 pm.
    b   Work out her pay for babysitting from 8 pm to 1 am.

4   Devon is paid a standard hourly rate, £x, plus £4 extra per hour for overtime.
    Write a formula for his pay, W, when he works n hours at his standard rate, plus m hours
    overtime.

5   Chris is 7 times as old as his daughter Bella.
    The sum of their ages is 40. How old is Chris?

    Q5 hint Use x for Bella's age.

Q5 A common error for this type of question is to use trial and error to find the values.
This can take much longer than using algebra, and there is more chance of making a mistake if you are
doing lots of calculations.

6   **Problem-solving** Tom is twice as old as Marie.
    Marie is 5 years older than Kay.
    The total of all their ages is 43.
    How old is Marie?

7   **Problem-solving** Rosie has 3 number cards.
    All the numbers are less than 30.
    The second number is 8 less than the first number.
    The third number is 4 times the second number.
    The sum of the 3 numbers is 50.
    What are the 3 numbers?

8   **Problem-solving** Here are a regular pentagon and a square.
    The sides of the square are 1 cm longer than the sides of
    the pentagon.
    The perimeter of the square is equal to the perimeter of
    the pentagon.
    Work out the perimeter of the square.

    l cm

9   Write a formula
    a   for the perimeter P of this shape
    b   for the area A of this shape.

    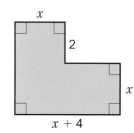

    x

    2

    x

    $x + 4$

10 Write a formula for the area A
    of this shape.

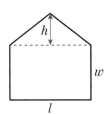

11 **Reasoning** Mitch and Ryan write a formula for the volume of this prism.
    Mitch writes $V = lwh$
    Ryan writes $V = \frac{1}{2}whl$

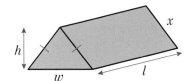

    a   Whose formula is correct?
        Show working to explain.
    b   Write a formula for the
        surface area of the prism.

Q11b hint Sketch a net.

2.2 Rearranging formulae

Objective

- Change the subject of a formula.

Key point 3

The **subject** of a formula is the letter on its own on one side of the equals sign.
You can change the subject of a formula using inverse operations.

1   Change the subject of each formula to the letter in brackets.
    a $y = mx$ $[x]$
    b $v = \frac{r}{t}$ $[r]$
    c $T = \frac{l}{w}$ $[w]$
    d $x - y = 2t$ $[t]$
    e $r - s = 5q$ $[s]$
    f $m + n = 6h$ $[h]$
    g $3x - y = 2z$ $[y]$
    h $3a + 2b = 5c$ $[c]$

2   Simplify
    a $-(2x + 3)$
    b $\frac{2}{-6}$
    c $\frac{x - y}{-1}$
    d $\frac{4x + 2}{-2}$

3   Make x the subject of each formula.
    a $z = y + 3x$
    b $3c = d + 2x$
    c $6e = 4x - 7$
    d $fg = h + 5x$
    e $kl = 3m - x$
    f $m = np - 2x$
    g $t = rs + px$
    h $uv = aw - xz$

4   Make the letter in brackets the subject.
    a $A = xyz$ $[y]$
    b $v = 5st$ $[t]$
    c $m = pqr$ $[q]$

    d $3s = 4tu$ $[u]$
    e $2xy = 8z$ $[x]$
    f $w = \frac{ab}{3}$ $[a]$

    g $d = \frac{ef}{5}$ $[f]$
    h $p = \frac{1}{2}rs$ $[r]$
    i $g = \frac{3hk}{5}$ $[k]$

    j $t = \frac{4ax}{7}$ $[a]$
    k $m = \frac{np}{x}$ $[n]$
    l $x = \frac{kt}{ab}$ $[b]$

Q4g A common error
is to divide both
sides by 5 instead of
multiplying.

5   Make y the subject.
    a $x = 2(y - z)$
    b $m = 5(t + y)$
    c $d = 4(k - y)$

    d $r = a(c + y)$
    e $n = \frac{5(y + z)}{3}$
    f $q = \frac{2(r - y)}{5}$

    g $x = \frac{k(y - n)}{6}$
    h $l = \frac{n(y - 1)}{4}$

Q5 hint Expand the
brackets first.

6 **Reasoning** Here are the steps for rearranging $x = \dfrac{y+5}{m} + 3$ to make y the subject.

A	B	C
$m(x-3) = y+5$	$y = m(x-3) - 5$	$x - 3 = \dfrac{y+5}{m}$

Put the steps in the correct order.

7 Make x the subject.

a $\quad t = \dfrac{x+2}{a}$

b $\quad v = \dfrac{x-3}{b}$

c $\quad n = \dfrac{x}{3} + m$

d $\quad y = \dfrac{x}{t} - d$

e $\quad z = \dfrac{x+4}{y} + 2$

f $\quad a = \dfrac{x-5}{b} + 3$

g $\quad h = \dfrac{x-k}{i} - 7$

h $\quad p = \dfrac{6-x}{r} + 4$

8 Make the letter in brackets the subject.

a $\quad x = by^2 \quad [y]$

b $\quad m = kz^2 \quad [z]$

c $\quad r = at^2 \quad [a]$

d $\quad c^2 = a^2 + b^2 \quad [b]$

e $\quad m = \sqrt{n} \quad [n]$

f $\quad t = \sqrt{2r} \quad [r]$

g $\quad R = ab^2c \quad [b]$

h $\quad s = \sqrt{\pi t} \quad [t]$

9 Make x the subject.

a $\quad x(5 - y) = z$

b $\quad 4x - xy = t$

c $\quad xr + 3x = m$

d $\quad 5x = s + tx$

e $\quad ax = b - 6x$

f $\quad cx + 4 = d - 2x$

g $\quad mx - 3 = t + nx$

h $\quad ux + s = rx - p$

> **Q9b hint** $4x - xy = x(\square - \square)$

> **Q9d hint** Rearrange so all the x terms are on one side.

10 Make the letter in brackets the subject.

a $\quad 2(5 - t) = 5t + a \quad [t]$

b $\quad 6(b - 4) = c + b \quad [b]$

c $\quad m(a - 3) = a + d \quad [a]$

d $\quad x + r = n(r + 4) \quad [r]$

e $\quad 2(3 + d) = 5(d - e) \quad [d]$

f $\quad 7(r - 1) = 4(p + r) \quad [r]$

g $\quad m(2 + x) = r(x - 1) \quad [x]$

h $\quad a(b + 3) = d(b + 5) \quad [b]$

11 **Reasoning** Match each formula to its rearrangement.

Formulae

a $\quad a^2 = b^2 + c^2$

b $\quad a = c + \sqrt{b}$

c $\quad a = \sqrt{b + c}$

d $\quad a = (b + 3)^2$

Rearrangements

A $\quad b = a^2 - c$

B $\quad b = (a - c)^2$

C $\quad b = \sqrt{a^2 - c^2}$

D $\quad b = \sqrt{a} - 3$

2.3 Expanding and factorising

Objectives

- Multiply double brackets.
- Factorise quadratic expressions.

Key point 4

To expand or multiply double brackets, multiply each term in one bracket by each term in the other bracket.

Key point 5

Expanding double brackets often gives a quadratic expression.

A quadratic expression always has a squared term (with a power of 2).

It may have a term with a power of 1 that is the same letter as the squared term.

It may also have a constant (number) term.

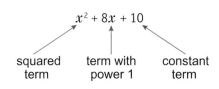

Key point 6

To square a single bracket, multiply it by itself, then expand and simplify.
$(x + 1)^2 = (x + 1)(x + 1)$

Key point 7

The **difference of two squares** is a quadratic expression with two squared terms, and one term is subtracted from the other.

For example $x^2 - 25$

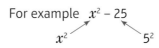

x^2 5^2

1   Copy and complete these expansions.
   a $(x + 5)(x + 2)$

×	x	$+ 2$
x	x^2	
$+ 5$		

   $= x^2 + \square x + \square x + \square$
   $= x^2 + \square x + \square$

   b $(x + 4)(x + 3) = x^2 + \square x + \square x + 12$ c $(x - 3)(x + 2) = x^2 + \square x - \square x - \square$
                        $= x^2 + \square x + 12$ $= x^2 - \square x - \square$

2   Which of these are quadratic expressions?
   a $x^2 + 5x + 4$ b $x^2 + 7$ c $x^3 + x^2 - 5$
   d $a^2 + a^4 + 1$ e $b^2 - 9$ f $8 + 4y + y^2$

3   Expand and simplify
   a $(x + 4)(x + 2)$ b $(x + 6)(x + 5)$ c $(a + 8)(a + 2)$
   d $(m + 9)(m + 6)$ e $(t + 7)(t + 5)$ f $(r + 4)(r + 7)$

4   Expand and simplify
   a $(x - 3)(x + 6)$ b $(x + 8)(x - 1)$ c $(y - 6)(y + 4)$
   d $(n - 7)(n + 1)$ e $(x - 4)(x + 9)$ f $(x - 3)(x + 11)$

5   Multiply each pair of linked terms.

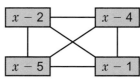

6   Square each expression.   **Q6a** A common error is to write $(x + 3)^2 = x^2 + 3^2$
   Simplify your answers.
   a $x + 3$ b $m - 2$ c $d + 9$ d $r + 10$
   e $x - 5$ f $y + 11$ g $z - 6$ h $x - 7$

7   Expand and simplify
   a $(x + 1)(x - 1)$ b $(x + 5)(x - 5)$ c $(x + 7)(x - 7)$ d $(x + 12)(x - 12)$

8   **Reasoning** Which of the quadratic expressions in **Q2** is a difference of two squares?

9   Factorise
   a $x^2 - 9$ b $x^2 - 36$ c $t^2 - 16$ d $h^2 - 100$

10 Reasoning Match equivalent expressions.

a $(x + 4)^2$ b $(x + 2)(x + 4)$ c $(x + 2)(x + 8)$ d $(x + 2)(x + 2)$

A $x^2 + 4x + 4$ **B** $x^2 + 10x + 16$ **C** $x^2 + 6x + 8$ **D** $x^2 + 8x + 16$

11 Factorise

a $x^2 + 13x + 42$ b $x^2 + 11x + 24$ c $x^2 + 6x + 5$

d $x^2 + 2x + 1$ e $y^2 + 9y + 14$ f $z^2 + 10z + 9$

12 Factorise

a $x^2 - 2x - 15$ b $x^2 + x - 20$ c $n^2 - 7n - 18$

d $t^2 - t - 56$ e $y^2 + 4y - 21$ f $x^2 - 4x - 12$

13 Factorise

a $x^2 - 9x + 14$ b $x^2 - 5x + 6$ c $x^2 - 12x + 32$

d $f^2 - 14f + 45$ e $v^2 - 18v + 80$ f $x^2 - 7x + 6$

14 Reasoning Match each quadratic expression to its factorisation.

a $x^2 - 64$ b $x^2 + 16x + 64$ c $x^2 - 16x + 64$ d $x^2 + 20x + 64$

A $(x - 8)^2$ **B** $(x + 4)(x + 16)$ **C** $(x - 8)(x + 8)$ **D** $(x + 8)^2$

15 Problem-solving Maya uses 20 of these parallelogram-shaped pieces of fabric to make patchwork.
Write an expression for the total area of the patchwork.

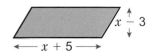

2.4 Inequalities

Objectives

- Solve linear inequalities.
- Show solutions to inequalities on a number line.

Key point 8

You can solve inequalities in the same way as linear equations.
You can solve two-sided inequalities using a balancing method.

Key point 9

When you multiply or divide both sides of an inequality by a negative number you reverse the inequality sign.

Warm up

1 Write down the inequalities represented on these number lines.

a 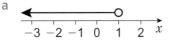   b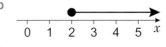

c 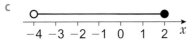   d

e 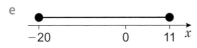   f

2 Write down the integer values of x that satisfy each inequality.

a $1 \leqslant x < 7$ b $0 < x \leqslant 3$

c $-2 < x < 4$ d $-2 \leqslant x \leqslant 1$

e $0 \leqslant x \leqslant 3.5$ f $-12.3 \leqslant x < -8$

Q2 communication hint An **integer** is a 'whole' number, e.g. –2, –1, 0, 1, 2, 3, …

3 Solve these inequalities.

a $x + 6 < 12$ b $x - 4 \geqslant 6$ c $x + 2 \geqslant -1$ d $x - 5 < 2$

e $-2 < x - 3$ f $3 \leqslant x + 7$ g $x - 4.5 \geqslant 3.2$ h $x + \frac{3}{4} < 2\frac{1}{4}$

4 Solve these inequalities.

a $5x \geqslant 20$ b $3x < 21$ c $8x \leqslant 16$ d $2x > -6$

e $6x < 9$ f $4x \geqslant 2$ g $14 < 7x$ h $1.5 > 3x$

5 Solve these inequalities.

a $\frac{x}{5} < 2$ b $\frac{x}{6} \leqslant 3$ c $\frac{x}{2} \geqslant 10$ d $\frac{x}{4} > 2.5$

e $\frac{x}{3} \leqslant \frac{1}{4}$ f $\frac{x}{10} > 0.2$ g $5 < \frac{x}{7}$ h $\frac{5}{6} \leqslant 10x$

6 Solve these inequalities.
Show each solution set on a number line.

a $2x - 1 > 9$ b $5x - 4 \leqslant 11$ c $\frac{x}{2} + 5 < 8$ d $4 + 3x \geqslant 19$

e $4x + 6 \leqslant -2$ f $2 + \frac{x}{3} < 5$ g $\frac{x}{4} - 2 \geqslant 5$ h $3x + 8 \geqslant -10$

7 Solve each inequality.

a $3(x + 5) > 24$ b $5(x - 2) \leqslant 20$ c $\frac{x + 1}{5} < 3$ d $\frac{x - 3}{2} \geqslant 5$

e $4(x + 2) > -8$ f $\frac{2 + x}{3} \leqslant 4$ g $\frac{2(3 + x)}{5} > 3$ h $\frac{3(x - 5)}{4} < 6$

8 **Reasoning** Match each inequality to a solution set.

a $3x + 7 < -2$ b $4(x - 1) > 8$ c $\frac{x}{5} + 2 < 1$ d $\frac{x + 5}{2} < 4$

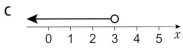 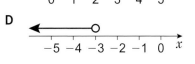

9 Solve these inequalities.

a $5x + 2 \leqslant 3x + 4$ b $3x + 9 > 2x - 6$ c $4x + 2 < 5x + 7$

d $3x - 6 \geqslant 4x - 5$ e $4x - 1 > 8x - 13$ f $2(x + 4) \leqslant 5x - 1$

g $12x + 3 < 10(x + 1)$ h $5(x - 2) > 3(x + 3)$

Q9 hint Collect all the x terms on the side with the larger number of xs.

10 **Problem-solving** I think of a positive integer, double it and add 4.
The answer is less than 11.
Find all the possible values for my number.

Q10 hint Use an inequality, not trial and error.

11 a Solve each inequality.

i $4 \leqslant 2x \leqslant 10$ ii $-10 \leqslant 5x < 5$ iii $-6 < 3x \leqslant 9$ iv $2 \leqslant 4x < 12$

v $-2 < 3x \leqslant 8$ vi $-1 \leqslant \frac{x}{5} \leqslant 1$ vii $0 \leqslant \frac{x}{8} < \frac{3}{4}$ viii $\frac{3}{8} < \frac{x}{2} \leqslant \frac{7}{10}$

b Show each solution set on a number line.

c Write all the possible integer values of x that satisfy each inequality.

12 Solve these double inequalities.

a $5 \leqslant x + 3 \leqslant 11$
b $-2 < x - 5 < 6$
c $-3 \leqslant 2x + 1 < 7$
d $1 \leqslant 3x - 2 \leqslant 10$
e $-1 < 5x + 4 < 6$
f $-6 \leqslant 2x - 3 \leqslant 3$

Q12a hint

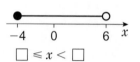

13 Solve

a $-4 < 2(x + 1) < 2$
b $-10 \leqslant 5(x - 1) < 5$
c $0 \leqslant 3(x + 7) \leqslant 20$
d $3 < 2(x - 5) \leqslant 6$
e $-1 \leqslant \dfrac{x + 3}{2} < 4$
f $-3 \leqslant \dfrac{x - 1}{4} < 1$

Q14a hint

$\times -1 \left(\begin{array}{c} -6 < -x \leqslant 4 \\ 6 > x \geqslant \square \end{array} \right) \times -1$

14 Solve these inequalities.

a $-6 < -x \leqslant 4$
b $-4 \leqslant -2x < 8$
c $-12 < -3x \leqslant 6$
d $0 < 4 - x < 3$
e $3 < 1 - 2x \leqslant 5$
f $-2 \leqslant 6 - 4x < 18$

$\square \leqslant x < \square$

15 **Problem-solving** Find the largest possible integer that satisfies $-15 < 2x + 3 \leqslant 0$

2.5 Sequences

Objectives

- Recognise and use Fibonacci-type sequences, quadratic sequences and geometric sequences.
- Find and use the nth term of an arithmetic sequence.

Key point 10

The **nth term** of a sequence tells you how to work out the term at position n (in any position).
It is also called the **general term** of the sequence.

Key point 11

In an **arithmetic sequence**, the terms increase (or decrease) by a fixed number called the **common difference**.
In a **Fibonacci-type sequence**, each term is the sum of the two terms before it.
In a **geometric sequence** the terms increase (or decrease) by a **constant multiplier**.
A **quadratic sequence** has n^2 and no higher power of n in its nth term.

1 For each sequence
   i find the missing terms   ii write the term-to-term rule.

a 3, 8, 13, $\square$, $\square$
b 14, 11, 8, $\square$, $\square$
c 6, $\square$, 10, $\square$, $\square$
d 4, $\square$, $\square$, 13, $\square$
e 1, -3, $\square$, $\square$, $\square$
f $\square$, -10, $\square$, 2, $\square$

2 Write down the next two terms in each geometric sequence
and the term-to-term rule.

a 5, 10, 20, $\square$, $\square$
b 1, 4, 16, $\square$, $\square$
c 800, 400, 200, $\square$, $\square$
d 0.003, 0.03, 0.3, $\square$, $\square$
e -625, -125, -25, $\square$, $\square$
f 3, -6, 12, $\square$, $\square$

Q2a hint

3 Write the next two terms in each Fibonacci-type sequence.

a 2, 2, 4, $\square$, $\square$
b -3, -3, -6, $\square$, $\square$
c 1, 2, 3, $\square$, $\square$
d -5, 2, -3, $\square$, $\square$
e 0.4, 0.4, 0.8, $\square$, $\square$
f 0.11, 0.11, 0.22, $\square$, $\square$

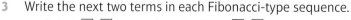

4   **Reasoning** Find the missing terms in these Fibonacci-type sequences.

   a   ☐, 5, 7, ☐, ☐   b   ☐, ☐, 4, 7, ☐   c   0, ☐, 6, ☐, 18
   d   ☐, ☐, 1, 1.5, ☐   e   5, ☐, 1, –3, ☐   f   ☐, 8, 2, ☐, ☐

5   **Problem-solving** The Fibonacci sequence starts
    1, 1, 2, 3, 5, …
   a   What is the first term greater than 50?
   b   Is 27 a term in this sequence? Explain how you know.

6   Write the first five terms of the sequences with these nth terms.

   a $2n$ b $n + 2$ c $4n - 2$ d $10n - 2$
   e $20 - 4n$ f $-3n + 1$ g $0.5n + 3$ h $5n + 6$

7   Write the 10th and 50th terms of the sequences with nth terms

   a $n - 4$ b $3n$ c $2n - 8$ d $5n + 1$
   e $100 - 4n$ f $-2n + 11$ g $\frac{n}{2} - 3$ h $0.1n + 5$

8   Find the nth term for each sequence.

   a   2, 9, 16, 23, 30, …   b   8, 11, 14, 17, 20, …
   c   5, 11, 17, 23, 29, …   d   9, 8, 7, 6, 5, …
   e   1, –1, –3, –5, –7, …   f   –7, –10, –13, –16, –19, …

   Q8a hint

   ☐n 7   14   ☐   ☐   ☐   ⎞ – ☐
          2   9   16   23   30
          +☐ +☐ +☐ +☐

9   **Problem-solving** What is the nth term for the
    sequence of odd numbers?

10 For each sequence, explain whether the number in brackets is a term in the sequence.

   a   –1, 2, 5, 8, 11, … (32)   b   8, 14, 20, 26, 32, … (102)
   c   2, –1, –4, –7, –10, … (86)   d   9, 16, 23, 30, 37, … (79)
   e   18, 14, 10, 6, 2, … (–40)   f   7, 11, 15, 19, 23, … (51)

   Q10 hint Work out
   the nth term.

11 Find the first term over 50 in each sequence.

   a   –1, 1, 3, 5, 7, …   b   –6, –3, 0, 3, 6, …
   c   –3, 1, 5, 9, 13, …   d   –5, 1, 7, 13, 19, …

   Q11 A common error is to spend a lot of time
   writing out the numbers in the sequence,
   instead of finding and using the nth term.

12 Find the term-to-term rule and the missing numbers in each of these geometric sequences.

   a   1, 3, 9, ☐, ☐   b   2, 10, ☐, 250, ☐   c   6, ☐, 24, ☐, 96   d   450, ☐, ☐, 0.45, 0.045

13 **Problem-solving** Penny decides to save £1 in January, £2 in February, £4 in March, and so on.
   a   In which month will she first save over £30?
   b   How much money will she then have saved in total?

14 Generate the first five terms of the sequences with these nth terms.

   a $2n^2$ b $5n^2$ c $n^2 + 1$ d $3n^2 - 1$ e $\frac{1}{2}n^2$ f $100 - n^2$

15 For each sequence
      i   work out the differences between
          consecutive terms
      ii describe the term-to-term rule.
   a   4, 7, 12, 19, 28
   b   15, 8, 3, 0, –1

   Q15a communication hint **Consecutive terms**
   are next to each other in the sequence.

   Q15a hint   4   7   12   19   28
               +☐ +☐ +☐ +☐

   Add the _____ numbers starting with ☐

16 **Reasoning** Match each sequence to its description.

   a   1, 4, 7, 10, 13, …   b   1, 4, 9, 16, 25, …   c   1, 4, 16, 64, 256   d   1, 4, 5, 9, 14, …

   A   Fibonacci-type   B   Geometric   C   Arithmetic   D   Quadratic

2.6 Straight-line graphs

Objectives

- Find the equation of a straight-line graph, and of a line joining two points.
- Interpret the gradient and y-intercept of a graph in context.

Key point 12

positive gradient   /   negative gradient   \   gradient of line = $\dfrac{\text{total distance up}}{\text{total distance across}}$

1   Write the equation of each line.

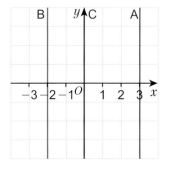

2   The graph shows bike hire costs from two different companies.
    Bikes 4 U charges £10 per hour.
    On Your Bike charges a £15 flat rate hire fee + £5 per hour.
    a   Match each line on the graph with the correct company.
    b   Matt wants a bike for 4 hours.
        Explain which company he should use.

3   Find the gradient of each line segment.

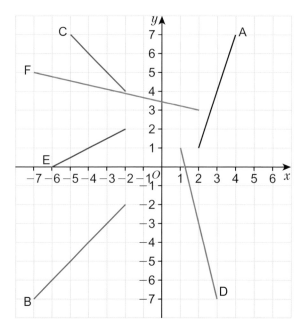

4   **Reasoning** Here are the equations of some graph lines.

$y = -x + 2$ $y = x + 3$ $y = \frac{1}{4}x - 1$ $y = 2x - 4$

$y = \frac{1}{4}$ $y = 5x + 1$ $y = 2x + 3$ $y = -2x + 4$

Which of these lines
a   are parallel
b   slope downwards from left to right
c   have the same y-intercept
d   is the steepest
e   has gradient zero?

5   Work out the equations of lines A, B, C, D and E.

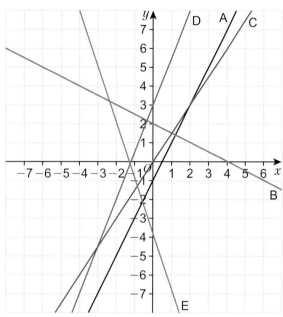

6   Jackie installs satellite dishes. She charges a £20 callout fee
    plus £25 per hour.
    a   Draw a graph of hours worked (x-axis) against total
        charge (y-axis).
    b   Work out the equation of the line.
    c   Explain what the gradient and y-intercept of the
        graph represent.

Q6a hint
Make a table of values.

Hours	0	1	2	3
Cost (£)				

7   Write the equation of each line.
    a   Gradient 5, passes through (0, 2)
    b   Gradient 7, passes through (1, 5)
    c   Gradient 0.5, passes through (3, 4)
    d   Gradient –2, passes through (5, 0)

Q7 hint $y = mx + c$

8   Draw these graphs from their equations.
    a $y - 2x - 3$ b $y = -\frac{1}{2}x + 2$ c $x + y = 7$ d $x + 2y = 6$

9   Work out the equation of the line that passes through
    each pair of points.
    a   (1, 0) and (4, 6)
    b   (–2, –5) and (0, 1)
    c   (4, –1) and (8, 1)
    d   (–1, 7) and (3, –1)
    e   (4, 6) and (–4, 4)
    f   (1, 0.5) and (1.5, 1.5)

Q9 hint Draw a graph or
substitute each pair of
coordinates into $y = mx + c$ and
use simultaneous equations.

2.7 Non-linear graphs

Objectives

• Draw quadratic, cubic and reciprocal graphs.
• Estimate solutions to an equation from its graph.

Key point 13

A quadratic function with a positive x^2 term has a symmetrical U-shaped curve called a **parabola**.

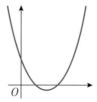

A quadratic function with a negative x^2 term has a symmetrical upside-down U-shaped curve.

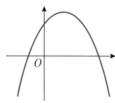

The curve always has a minimum or maximum turning point.

Key point 14

To solve the equation $ax^2 + bx + c = 0$, read the x-coordinates where the graph crosses the x-axis. The values of x that satisfy the equation are called **roots**.

Key point 15

A cubic function can have one, two or three roots.

Warm up

1 Write down
   a the equation of the line of symmetry
   b the y-intercept
   c the turning point
   of this graph.

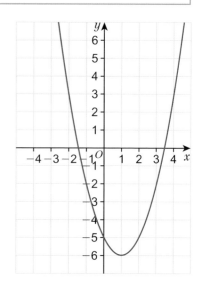

2 a Copy and complete the table of values for $y = x^2 + 3$.

x	−3	−2	−1	0	1	2	3
y	12					7	

   b Plot the graph of $y = x^2 + 3$.
   c Write the equation of the line of symmetry.
   d What is the turning point?
   e What is the y-intercept?

Q2 A common error is to not draw smooth curves, or to use a blunt pencil so the graph is inaccurate.

3 Draw the graphs of
   a $y = x^2 - 2$
   b $y = x^2 + 4$
   for $-3 \leqslant x \leqslant 3$.

Q3 hint Use a table of values like the one in **Q2**.

4 a Copy and complete the table of values for $y = x^2 - 3x + 1$.

x	-2	-1	0	1	2	3	4
y							

Q4 hint You could add rows for x^2, $3x$ and +1 to the table.

   b Draw the graph of $y = x^2 - 3x + 1$.
   c Write down
      i   the equation of the line of symmetry
      ii the coordinates of the turning point.
   d Use your graph to estimate the solutions to $x^2 - 3x + 1 = 0$.

5 a On graph paper, draw the graph of $y = x^2 - 2x - 3$ for $-2 \leqslant x \leqslant 4$.
   b By drawing suitable lines on your graph, estimate the solutions to
      i $x^2 - 2x - 3 = 0$
      ii $x^2 - 2x - 3 = 4$
      iii $x^2 - 2x - 3 = -1$
      iv $x^2 - 2x = -1$

Q5b iv hint Rearrange to $x^2 - 2x - 3 = \square$

6 a Copy and complete the table of values for $y = x^3 - 3$ between $x = -3$ and $x = 3$.

x	-3	-2	-1	0	1	2	3
y	-30						

   b Draw the graph of $y = x^3 - 3$ for $-3 \leqslant x \leqslant 3$.
   c Estimate the solution to
      i $x^3 - 3 = 0$
      ii $x^3 - 3 = 8$

7 Draw the graph of $y = x^3 - 2x$
  for $-3 \leqslant x \leqslant 3$.

Q7 hint Make a table of values like the one in **Q6**.

8 **Reasoning** a Copy and complete the table of values for $y = -\dfrac{1}{x}$.

x	-4	-3	-2	-1	$-\frac{1}{2}$	$-\frac{1}{4}$	$\frac{1}{4}$	$\frac{1}{2}$	1	2	3	4
$-\dfrac{1}{x}$	$\frac{1}{4}$			1				-2				

   b Draw the graph of $y = -\dfrac{1}{x}$.
   c What is same and what is different
      about the graphs of $y = \dfrac{1}{x}$ and $y = -\dfrac{1}{x}$?

Q8c hint Sketch the graph of $y = \dfrac{1}{x}$ to compare.

9 a Copy and complete the table of values for $y = \dfrac{2}{x}$.

x	-4	-3	-2	-1	$-\frac{1}{2}$	$-\frac{1}{4}$	$\frac{1}{4}$	$\frac{1}{2}$	1	2	3	4
$\dfrac{2}{x}$	$-\frac{1}{2}$		-1					4				

   b Draw the graph of $y = \dfrac{2}{x}$, where $x \neq 0$, for $-4 \leqslant x \leqslant 4$.
   c Use your graph to estimate
      i   the value of y when $x = 0.6$
      ii the value of x when $y = -7.5$

2.8 Solving equations

Objectives

- Solve quadratic equations algebraically.
- Solve simultaneous equations using graphs or algebra.

Key point 16

To solve a quadratic equation:
- rearrange if necessary so the quadratic expression is equal to zero.
- factorise the expression.

Key point 17

Simultaneous equations are equations that are both true for a pair of variables (letters).
To find the solution to simultaneous equations:
- draw the lines on a coordinate grid
- find the point where the lines cross (the point of **intersection**).

Key point 18

To solve simultaneous equations by the elimination method, add or subtract the equations to eliminate either the x or the y terms.
You may need to multiply one or both equations by a number.

1 Draw the graphs of $2y + x = 6$ and $y - 2x = -2$ on the same axes.

2 Factorise
   a $x^2 - 100$ b $x^2 + 4x + 3$
   c $x^2 - 2x - 35$ d $x^2 - 7x + 12$

   Q2 hint $(x\ \Box)\ (x\ \Box)$

3 Solve these quadratic equations.
   a $x^2 - 49 = 0$ b $x^2 - 81 = 0$ c $x^2 - 9 = 0$
   d $x^2 - 121 = 0$ e $x^2 - 64 = 0$ f $x^2 - 144 = 0$

   Q3 A common error is to write only the positive solution to a quadratic equation like this.

4 Solve
   a $x^2 - 12 = 13$ b $x^2 + 3 = 19$ c $x^2 + 1 = 170$
   d $-x^2 + 4 = 0$ e $36 - x^2 = 0$ f $x^2 + 1.75 = 2$

   Q4 hint Rearrange to get $x^2 - \Box = 0$ first.

5 Solve
   a $x^2 + 8x + 15 = 0$ b $x^2 + 4x - 12 = 0$ c $x^2 + 4x - 21 = 0$
   d $x^2 + 18x + 81 = 0$ e $x^2 + 4x - 32 = 0$ f $x^2 - 9x + 18 = 0$
   g $x^2 - 5x + 6 = 0$ h $x^2 - 8x + 16 = 0$ i $x^2 - 3x - 40 = 0$

6 **Reasoning** Maisie says that $x^2 - 10x + 25 = 0$ and $x^2 + 10x + 25 = 0$ have the same solutions.
   Show working to explain why she is wrong.

7 Solve
   a $x^2 = 3x + 18$ b $x^2 + 8x = -7$ c $8 = 9x - x^2$
   d $x^2 + 12x + 35 = 0$ e $x^2 - 10x + 40 = 16$ f $x^2 - x - 15 = -3$

   Q7 hint Rearrange to get $x^2 + \Box x + \Box = 0$ first.

8 **Problem-solving** The area of this rectangle is $16\,\text{cm}^2$.

   $x - 6$
   x

   Q8 strategy hint Write and solve an equation.

   Find the length and the width.

9 Solve

    a $2x^2 - 200 = 0$ b $3x^2 - 48 = 0$ c $5x^2 - 45 = 0$

    d $4x^2 - 64 = 0$ e $6x^2 - 24 = 0$ f $3x^2 - 75 = 0$

> **Q9 hint** Divide by the common factor first.

10 Use the graphs you drew in **Q1** to solve the simultaneous equations $2y + x = 6$ and $y - 2x = -2$.

11 Draw graphs to solve these pairs of simultaneous equations.

    a $x + y = 4$ and $y + 3x = 6$

    b $y - 3x = 1$ and $3y + x = 3$

    c $y - 2x = 4$ and $2y + x = 3$

12 Solve these simultaneous equations algebraically.

    a $2x + 3y = 11$ b $x + 2y = 10$ c $4x + 2y = -8$

       $5x + 3y = 14$ $x - 3y = 0$ $2x - 3y = -12$

    d $2x - 3y = 7$ e $3x + 2y = -16$ f $2x - 3y = -22$

       $5x + y = 9$ $2x - 5y = 21$ $5x - 4y = -20$

13 Solve these simultaneous equations algebraically.

    a $3x + 4y = 6$ b $4x - 3y = 12.3$

       $x + 4y = 3.6$ $3x - 2y = 8.7$

14 **Problem-solving** 2 pizzas and an ice cream cost £16.10

    3 pizzas and 4 ice creams cost £29.65

    Work out the cost of

    a 1 pizza   b 1 ice cream.

2.9 Proof

Objectives

- Recognise equations and identities.
- Prove results using algebra.

> **Key point 19**
>
> An **equation** has an equals sign. You can solve it to find the value of the letter.
> An **identity** is similar, but is true for all values of x and uses the symbol '$\equiv$'.

> **Key point 20**
>
> To show a statement is an identity, expand and simplify the expressions on one or both sides of the identity sign, until the two expressions are the same.

> **Key point 21**
>
> An even number is a multiple of 2.
> $2m$ and $2n$ are both general terms for even numbers where m and n are integers.

1 Write expressions in terms of n for the numbers in this grid.

$n = 5$	6
12	13

2 Expand

    a $(x + 2)(x - 3)$ b $(x - 7)(x + 2)$ c $(x - 5)(x - 3)$

3   Decide if these are equations or identities.
   a $x^2 + 2x + 1 = 0$
   b $4x^2 - 2x = 2x(2x - 1)$
   c $(x - 5)^2 + 2 = x^2 - 10x + 27$
   d $x^2 - 1 = 20 + 4x$

4   Show that
   a $5x^2 - 3x^3 \equiv x^2(5 - 3x)$
   b $(x + 7)(x - 7) \equiv x^2 - 49$
   c $(x + 3)(x + 4) + (x + 4) \equiv (x + 4)^2$
   d $(x + 6)^2 - 6x \equiv (x + 3)^2 + 27$
   e $(x + a)(x + b) \equiv x^2 + (a + b)x + ab$
   f $x(x + 2) \equiv (x + 1)^2 - 1$

5   Show that the area of this shape is $x^2 + 5x - 9$.

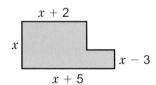

6   Show that the area of this triangle is $x^2 + 3x$.

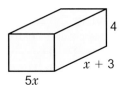

7   **Problem-solving** Show that the volume of the cuboid is 4 times the volume of the triangular prism.

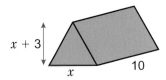

8   In this grid, a 2 × 2 square is highlighted.

1	2	3	4	5
6	7	8	9	10
11	12	13	14	15
16	17	18	19	20
21	22	23	24	25

   a   Show that the sum of 13 + 14 is 10 more than the sum of 8 and 9.
   b   Choose another 2 × 2 square and work out
       (sum of the two larger numbers) − (sum of the two smaller numbers)
   c   Use algebra to show that this is true for all the 2 × 2 squares in the grid.

9   **Problem-solving** Show that for any 2 × 2 square in this grid, the products of the diagonals have a difference of 6.

1	2	3	4	5	6
7	8	9	10	11	12
13	14	15	16	17	18
19	20	21	22	23	24

Q9 communication hint To find the **product** of two numbers, multiply them together.

10 **Reasoning** Look back at the grid in **Q8**.
   a   Predict the difference between the products of the diagonals in a 2 × 2 square.
   b   Use algebra to show if your prediction is correct.

2.10 Mixed exercise

Objective

• Consolidate your learning with more practice.

1 This square has side length x and area $17\,\text{cm}^2$.
   a Write an equation in x for the area of the square.
   b Solve to find the side length.
     Give your answer in surd form.

2 I think of a positive number, square it and add 14. My answer is 95.
   a Write an equation using n for the mystery number.
   b Solve it to find n.

3 **Exam question**

   a Simplify $4b \times 2c$ **(1 mark)**
   b Expand $3(2w - 5t)$ **(2 marks)**
   c Expand and simplify $(x + 7)(x - 2)$ **(2 marks)**
   November 2010, Q2, 5MB2H/01

   Exam hint
   In part **b** some students incorrectly tried to simplify by adding terms with different letters. The most common error in part **c** was to forget the negative sign before the 2 or to make mistakes in multiplying positive and negative numbers.

4 The blue card is a rectangle with length $x + 4$ and width $2x$.
   A rectangle with length $2x - 1$ and width x
   is cut out of the blue card.
   Show that the area of the remaining card is $9x$.

   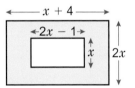

5 **Exam question**

   Here are the first 5 terms of an arithmetic sequence.

   3   9   15   21   27

   a Find an expression, in terms of n, for the nth term of this sequence. **(2 marks)**

   Ben says that 150 is in the sequence.

   b Is Ben right? You must explain your answer. **(1 mark)**
   February 2013, Q5, 1MA0/1H

   Exam hint
   In part **b**, writing 'Yes' or 'No', without showing any working to explain why, does not get any marks.

6 The yellow card is a rectangle with length $x + 5$ and width $x + 4$.
   A rectangle with length $x + 2$ and width x is cut out of the
   yellow card. The area of the remaining card is $41\,\text{cm}^2$.
   Find x.

   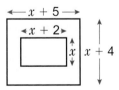

7 **Exam question**

   $-2 < n \leqslant 3$

   a Represent this inequality on the number line.

   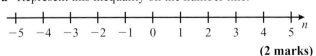

   (2 marks)

   b Solve the inequality $8x - 3 \geqslant 6x + 4$ **(2 marks)**
   June 2013, Q14, 1MA0/1H

   Exam hint
   In part **a**, some students showed the correct end points of the interval, but did not join them with a line. In part **b**, a common error was to write the answer as $x = \square$, instead of $x > \square$.

8 Solve these simultaneous equations algebraically.
   $2x - y = 1$
   $x + y = 8$

9 Pia says that $(x + 5)^2 = x^2 + 25$.
   Explain what she has done wrong.

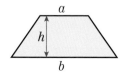

Q9 hint Pia has made a common error – make sure you don't make it.

10 Solve these simultaneous equations.
    $x + 3y = 12$
    $2x - y = 10$

11 a Write the formula for the area, A, of this trapezium.

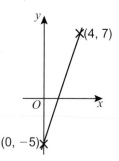

    b Rearrange your formula to make h the subject.
    c Rearrange your formula to make b the subject.

12 A straight line passes through $(0, -5)$ and $(4, 7)$.
    Find the equation of the straight line.

Q12 hint There are at least two methods you could use for this question. Whichever method you choose, show your working clearly.

13 Is 7 a possible solution to the inequality
    $7 \leqslant 2x + 5 \leqslant 20$?
    Show how you got your answer.

14

Exam question

    a Simplify $(m^{-2})^5$ **(1 mark)**
    b Factorise $x^2 + 3x - 10$ **(2 marks)**

    June 2012, Q16, 1MA0/1H

Exam hint
In part **b** a common error was to factorise the two terms to $x(x + 3) - 10$.

15 Find the integer value of x that satisfies both the inequalities
    $5x - 8 < 22$ and $2x + 1 > 9$

16 a Copy and complete the table of values for $y = \dfrac{4}{x}$

x	0.5	1	2	4	5	8
y		4	2			

Q16 A common error in part **b** is to join the points with straight lines, instead of a smooth curve.

    b Draw the graph of $y = \dfrac{4}{x}$ for $0.5 < x < 8$

3 RATIO, PROPORTION AND RATES OF CHANGE

What is covered in this section	Foundation Student Book reference
3.1 Calculations • Solve problems using ratios.	Unit 11
3.2 Growth and decay • Calculate percentage change. • Find the original amount after a percentage change. • Solve growth and decay problems.	Unit 14
3.3 Compound measures • Solve problems involving density and pressure.	Unit 14
3.4 Rates of change • Draw and interpret real-life graphs showing changes in quantities over time.	Units 9, 11
3.5 Direct and inverse proportion • Solve problems involving direct and inverse proportion.	Units 11, 14, 20
3.6 Similarity • Find missing lengths and angles in similar shapes.	Unit 19
3.7 Mixed exercise • Consolidate your learning with more practice.	

3.1 Calculations

Objective

• Solve problems using ratios.

> **Key point 1**
>
> You can compare ratios by writing them as **unit ratios**. In a unit ratio, one of the numbers is 1.

1 Share
   a £50 in the ratio 1:4
   b 6.6 litres in the ratio 1:5
   c 700 g in the ratio 5:3
   d £96 in the ratio 3:7

2 Megan and Tom share some money in the ratio 3:5.
   Megan gets £30 less than Tom.
   How much in total did they share between them?

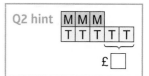

3 Sunil and Asha share some money in the ratio 2:7.
   Asha has £35 more than Sunil.
   How much does Asha have?

4 **Problem-solving** Bill and Trevor buy a painting for £12 000.
   Bill pays £7500 and Trevor pays the rest.
   After 10 years they sell it for £10 250.
   How should they share the money fairly?

5   Anna, Bob and Carly share a packet of fudge in the ratio $2:3:5$. Anna has 150 g less than Carly.
    a   How much does Carly have?   b   How much does Bob have?

6   In a set of counters, $\frac{2}{5}$ of the counters are white, $\frac{1}{5}$ are blue and the remainder are red.
    Write the ratio of
    a   red to white to blue counters   b   blue to red counters.   Q6 hint Draw a diagram.

7   **Reasoning** Three numbers are in the ratio $1:5:7$.
    The difference between the largest and smallest is 36.
    Find the three numbers.

8   Sunset orange paint is mixed from red, yellow and white paint in the ratio $13:16:2$.
    Tropical orange paint is mixed from red, yellow and white paint in the ratio $8:9:1$.
    a   Which colour has the higher proportion of yellow paint?
    b   Tariq mixes 240 ml red, 270 ml yellow and 30 ml white paint. Which colour is he making?

9   Write each of these in the form $n:1$.
    a $37.5:2.5$ b $90.2:8.2$ c $30\,\text{m}:50\,\text{cm}$
    d   £19.50 : £4.25   e   4 hours : 20 minutes   f $5\,\text{kg}:320\,\text{g}$
    g $\frac{3}{8}:\frac{1}{6}$ h $9.4:0.2$

    Q9 A common error is to include units of measure in unit ratios.

10 Concrete is made using cement, sand and gravel in the ratio $1:3:5$.
    a   Jim has 138 kg of gravel. How much sand and cement does he need?
    b   Marnie has $4\frac{1}{2}$ kg of sand. How much cement and gravel does she need?

11 Shireen and Mike share some money in the ratio $2:3$.
    Write the missing fractions.   Q11 hint Draw a diagram.

    a   Shireen's amount is $\frac{\square}{\square}$ of Mike's amount.   b   Mike's amount is $\frac{\square}{\square}$ of Shireen's amount.

12 There are two orange squash mixes.
    • Squash A is $\frac{1}{7}$ squash, remainder water   • Squash B is $\frac{15}{17}$ water, remainder squash
    Write the ratio of squash to water for each.

13 a   Copy and complete the table of values for $y = 3x$.
    b   Write the ratio $y:x$ for each pair of values in its
        simplest form. What do you notice?

x	1	2	3	4
y				

    c   Write the ratio $x:y$ in its simplest form.
    d $s = 5t$
        i   Write the ratio $s:t$ ii   Write the ratio $t:s$
    e $y = \dfrac{x}{2}$. Write the ratio $y:x$ in its simplest form.

3.2 Growth and decay

Objectives
•   Calculate percentage change.
•   Find the original amount after a percentage change.
•   Solve growth and decay problems.

Key point 2

You can calculate a **percentage change** using the formula

$$\text{percentage change} = \frac{\text{actual change}}{\text{original amount}} \times 100$$

Key point 3

Banks and building societies pay **compound interest**. At the end of the first year, interest is paid on the money in the account. The interest is added to the amount in the account.
At the end of the second year, interest is paid on the original amount in the account *and* on the interest earned in the first year, and so on.

1   Increase
    a   £45 by 20%   b   975 km by 2%

2   Decrease
    a   £75 by 15%   b   5 kg by 3%

3   Celine bought a chair for £12 and then sold it at an auction for £22.50.

> **Q3** A common error with percentage profit and loss questions is to use the new price instead of the actual change in price.

    a   What was her actual profit?
    b   What was her percentage profit?

4   In 1977 there were only 1100 giant pandas alive in the wild.
    In 2014 there were 1864.
    What was the percentage increase in the giant panda population?
    Give your answer to the nearest 1%.

5   **Problem-solving** The UK 2011 census recorded 3.5 million children under 5 in England and Wales. This was 406 000 more than in 2001.
    Calculate the percentage increase in the number of children under 5 from 2001 to 2011.

6   Rachael bought her prom dress for £180.
    After the prom she sold it for £72.
    What was the percentage loss?

7   All these prices include VAT.
    Work out the price of each item excluding VAT.

> **Q7 hint** VAT is 20%.

    a   laptop £330   b   smartphone £132
    c   chocolate bar 60p   d   take away meal £10.20

8   **Problem-solving** Tom bought a coat for £49 in a '30% off' sale.
    How much money did he save?

9   In a sale the price of a washing machine is reduced by 15%.
    On Saturday there is a further 5% off.
    What is the price on Saturday of a washing machine that originally cost £499?

> **Q9 hint** On Saturday, another 5% is taken off the '15% off' price.

10 In 2005 a house was valued at £190 000.
    From 2005 to 2010 house values fell by 8%.
    From 2010 to 2015 they rose by 12%.
    What was the value of the house in 2015?

11 Work out the total amount in each account at the end of 2 years.
    a   £85, compound interest rate 1.9% p.a.
    b   £5000, compound interest rate 2.2% p.a.
    c   £2450, compound interest rate 1.67% p.a.
    d   £250 000, compound interest rate 0.35% p.a.

> **Q11 communication hint p.a.** means 'per annum' or 'per year'.

3.3 Compound measures

Objective

- Solve problems involving density and pressure.

Key point 4

Density is a compound measure. It is the **mass** of the substance contained in a certain **volume**. To calculate density in g/cm³, you need to know the mass in grams (g) and the volume in cubic centimetres (cm³).

$$density = \frac{mass}{volume} \quad or \quad D = \frac{M}{V}$$

Density is usually measured in g/cm³. Density can also be measured in kg/m³.

Key point 5

Pressure is a compound measure. It is the **force** applied over an **area**. To calculate pressure, you need to know the force in newtons (N) and the area in square metres (m²).

$$pressure = \frac{force}{area} \quad or \quad P = \frac{F}{A}$$

Pressure is usually measured in N/m².

Warm up

1   A piece of silver with volume 20 cm³ has mass 210 g.
    Work out its density in g/cm³.

2   A force of 2.4 N is applied over an area of 3 m².
    Work out the pressure in N/m².

3   A gold ingot of volume 80 cm³ weighs 1554 g.
    Work out its density in g/cm³.

4   A steel bar of volume 1270 cm³ has mass 10 kg.
    Calculate its density in g/cm³.

    Q4 hint For density in g/cm³ you need mass in g and volume in cm³.

5   A titanium pendant with volume 60 mm³ has mass 0.27 g.
    Calculate the density of titanium in g/cm³.

6   The density of lead is 11.3 g/cm³.
    Find the mass of 50 cm³ of lead.

    Q6 hint Substitute the values into the formula to get an equation to solve.

7   The density of nickel is 8.9 g/cm³.
    Calculate the volume of 750 g of nickel, to the nearest cm³.

8   A plastic block weighing 600 N with base area 150 cm² stands on the floor.
    What pressure does the block exert on the floor? Give your answer in N/cm².

9   The pressure of the water on the bottom of a pond is 1900 N/m².
    The area of the bottom of the pond is 2.5 m².
    What force does the water exert?

10 A force of 175 N is applied to a piston. The pressure is 6.25 N/cm².
    Calculate the area of the piston in cm².

11 **Reasoning** A woman of mass 65 kg exerts a force of 650 N on the floor.
    The soles of her high-heeled shoes have a total area of 180 cm².
    A man of mass 85 kg exerts a force of 850 N on the floor.
    The soles of his trainers have a total area of 320 cm².
    Who exerts the greater pressure on the floor? Show working to explain.

3.4 Rates of change

Objective

- Draw and interpret real-life graphs showing changes in quantities over time.

Key point 6

A **rate of change graph** shows how a quantity changes over time.
The **gradient** of a straight-line graph is the **rate of change**.

Key point 7

On a **distance–time graph** the gradient represents the speed.

Average speed $= \dfrac{\text{total distance}}{\text{total time}}$

On a **velocity–time graph** the gradient represents the acceleration.

Acceleration $= \dfrac{\text{change in velocity}}{\text{time}}$

Key point 8

When two quantities are in **direct proportion**, plotting them as a graph gives a straight line through the origin. The origin is the point (0, 0) on a graph.

1   The graph shows how the depth of water in a reservoir changes over time.
    a   Write down the y-intercept of this graph.
    b   What does the y-intercept represent?
    c   Find the gradient.
    d   What does the gradient represent?
    e   Write the equation of the line, $d = \square\, t + \square$

> **Q1** A common error is to not read the scales on the axes carefully.

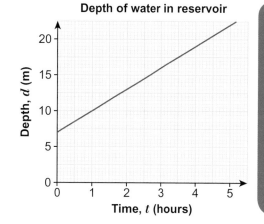

2   Train A travels from London to Penzance.
    Train B travels from Penzance to London.
    Use the graph to estimate
    a   the time the train left Penzance
    b   the time when they pass each other

> **Q2b hint** They pass each other when they are at the same place at the same time.

    c   their distance from London when they pass each other.
    d   When was train A travelling fastest? How can you tell this from the graph?
    e   Work out the average speed for each train.
    f   Which train travelled faster on average?

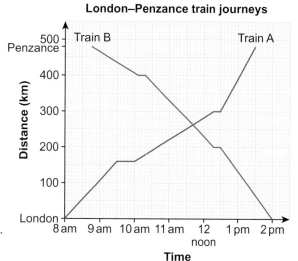

3 The timetable is for a train journey from Banbury to Haddenham and Thame Parkway.

Station	time
Banbury (depart)	1313
King's Sutton (arrive)	1318
Bicester North (arrive)	1330
Haddenham and Thame Parkway (arrive)	1343

> **Q3** A common error is to use the wrong units. To calculate speeds in km/h, use distance in km and time in hours in the formula. Don't leave the time in minutes.

The train stops for 3 minutes at each station. The distances between the stations are shown here.

Banbury 6 km King's Sutton 19 km Bicester North 24 km Haddenham and Thame Parkway

a Draw a distance–time graph for this journey.

> **Q3a hint** Start your 'time' axis at 1310.

b Work out the average speed of the train for each stage of the journey. Give your answer in km/h.

c Work out the average speed for the whole journey.

4 The graph shows the volume of water in a leaking paddling pool.

a How much water was in the paddling pool at the start?

b Find the gradient of the graph.
What does this represent?

c Estimate how long it will take for the paddling pool to empty completely.

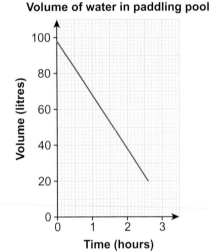

5 **Reasoning** Here is a table of values for x and y.

x	2	3	4	6
y	5	7.5	10	15

a Draw the graph for this table of values.

b Work out the equation of the line.

c Are x and y in direct proportion? Explain how you know.

> **Q5c hint** Extend your line. Does it pass through (0, 0)?

6 **Problem-solving** Here is a table of values for t and v.

t	1	2	4	7
v	5	7	11	17

a Write a formula connecting t and v.

b Are t and v in direct proportion? Explain how you know.

7 **Reasoning** The graph shows the temperature inside a freezer.

a What does the y-intercept tell you?

b The freezer warmed up during a power cut.
When did the power come back on?
Explain how you know.

c Estimate how long the temperature was above –10 °C.

d Estimate the temperature at 3 pm.

e Predict the temperature at 4.30 pm.

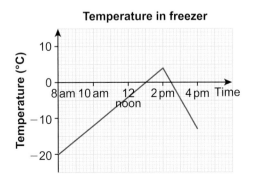

8 The graph shows the motion of some cars.
   a What was the initial velocity of car A?

   > Q8a communication hint **Initial velocity** means 'starting velocity'.

   b How long did it take car A to reach 60 km/h?
   c Work out the acceleration of car A
      from 0 to 60 km/h.
      Give your answer in km/h².

   > Q8c hint Find the gradient.

   d Find the initial velocity and acceleration for
      i car B   ii car C.

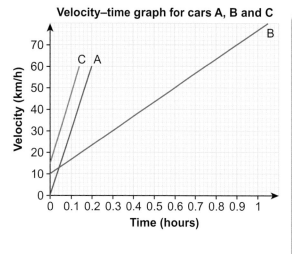

Velocity–time graph for cars A, B and C

9 **Problem-solving** The table shows how the percentage of the adult population who smoked changed between 1974 and 2013.

Year	Percentage of adults in England who smoked
1974	46%
1980	39%
2002	26%
2009	21%
2013	19%

From this data, estimate
a when the percentage first fell below 20%
b when the percentage of smokers will be below 10%.
c Which of your answers do you think is more reliable? Why?

> Q9 hint You could draw a graph for the data.

3.5 Direct and inverse proportion

Objective

• Solve problems involving direct and inverse proportion.

Key point 9

When
• y varies as x
• y varies directly as x
• y is in direct proportion to x
you can write $y \propto x$
$y \propto x$ means 'y is proportional to x'.
When $y \propto x$, then $y = kx$, where k is the **constant of proportionality**.

Key point 10

When two variables x and y are in inverse proportion,

$$X \propto \frac{1}{Y} \qquad X = \frac{k}{Y} \qquad Y = \frac{k}{X}$$

$XY = k$ (constant)

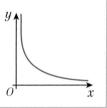

Key point 11

When x and y are in direct proportion, $y = kx$.

When x and y are inversely proportional, $y = \frac{k}{x}$

1 Solve
   a $24 = 3k$
   b $8 = \frac{k}{2}$
   c $5.4 = 2k$
   d $1.6 = \frac{k}{3}$

2 Write each statement using k, the constant of proportionality.
   a $F \propto a$
   b $v \propto \frac{1}{t}$
   c $x \propto \frac{1}{y}$
   d $c \propto d$

3 y is proportional to x.
   When $x = 3$, $y = 12$.
   a Use the given values of x and y to find the constant of proportionality, k.
   b Write a formula connecting y and x.
   c Find y when $x = 1$.
   d Find x when $y = 10$.

   Q3b hint Use your value of k.

4 A is directly proportional to m.
   When $A = 2.4$, $m = 6$.
   a Write a formula for A in terms of m.
   b Find m when $A = 5$.

5 **Problem-solving** s is directly proportional to d.
   When $d = 25$, $s = 50$.
   Find s when $d = 11.2$.

6 Two electricians take 3 days to rewire a house. How long will it take
   a one electrician to rewire one house
   b 4 electricians to rewire one house
   c 8 electricians to rewire 2 houses?

7 **Problem-solving** Four tractors plough 250 hectares in 6 hours.
   How long would it take with 3 tractors?

8 r is inversely proportional to t.
   When $r = 2$, $t = 5$.
   a Write a formula for r in terms of t.
   b Find r when $t = 20$.

   Q8 A common error is to not read the question carefully. Is it inverse or direct proportion?

9 **Problem-solving** d is inversely proportional to v.
   $d = 1.6$ when $v = 0.5$. Find v when $d = 7.2$.

10 A factory has a large order for cotton fabric. The table shows the time it will take to make the fabric with different numbers of machines.
   a Draw a graph with **Number of machines** on the x-axis and **Time** on the y-axis.
   b Describe the relationship between the number of machines and the time taken.
   c Estimate how long it would take 5 machines to make the fabric.

Number of machines	Time (hours)
2	12
3	8
4	6
6	4
8	3

Q10b hint What type of proportion?

3.6 Similarity

Objective
* Find missing lengths and angles in similar shapes.

Key point 12

These triangles are **similar**.
Corresponding sides are shown in the same colour.
Corresponding angles are shown in the same colour.

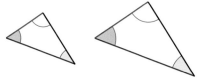

Key point 13

For similar shapes:
* **corresponding sides** are all in the same ratio.
* **corresponding angles** are equal.

Key point 14

When a shape is enlarged, the perimeter of the shape is enlarged by the same **scale factor**.

---

1   These two rectangles are similar.
   a   Find the height of the small rectangle.
   b   Find the length of the diagonal of the large rectangle.

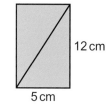

6.5 cm   12 cm
2.5 cm   5 cm

2   Triangle ABC is similar to triangle DEF.

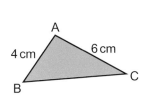   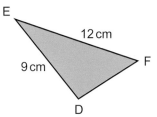

4 cm   6 cm
9 cm   12 cm

   a   Which side corresponds to
      i AB   **ii** ED   **iii** BC?

   > **Q2a hint** Sketch the triangles the same way up.

   b   What is the scale factor of the enlargement that maps
      i   triangle ABC to triangle DEF   **ii**   triangle DEF to triangle ABC?
   c   Work out the length of DF.
   d   Work out the length of BC.

3   Triangles PQR and XYZ are similar.

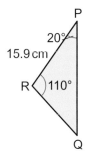   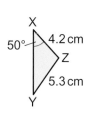

20°   50°   4.2 cm
15.9 cm   5.3 cm
110°

   a   Work out the length of RQ.
   b   Work out the size of
      i   angle XYZ   **ii**   angle RQP.

4 **Reasoning** a Explain how you know that these two quadrilaterals are similar.

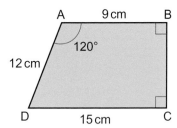 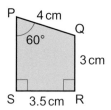

   b Work out the length of BC.   c Work out the length of PS.

5 **Problem-solving** These two quadrilaterals are mathematically similar.
Find the missing lengths x and y.

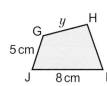 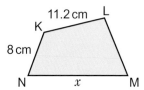

6 The lines BC and DE are parallel. Triangles ABC and ADE are similar.

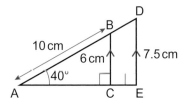

   a Write down the size of angle ABC.   b Work out the length of AD.
   c Work out the length of BD.

7 **Reasoning** The lines GJ and HI are parallel.
   a Explain why triangles FJG and FIH are similar.
   b Work out the length of IJ.

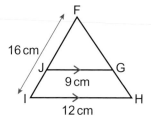

> **Q7a hint** Label equal angles with the same letter.

8 **Problem-solving** The lines PQ and RS are parallel.
RT = 4 cm   TS = 5 cm

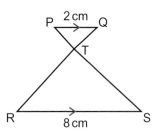

> **Q8** A common error in similarity problems is to match up the pairs of corresponding sides incorrectly.

Work out the length of
   a QT   b PT

9 **Reasoning** These two shapes are mathematically similar.
The perimeter of shape B is 18 cm.
Work out the perimeter of shape A.

3.7 Mixed exercise

Objective

- Consolidate your learning with more practice.

1   A drink is made from cordial, fruit juice and lemonade in the ratio 2:3:7.
    a   What fraction of the drink is fruit juice?
    b   How many litre bottles of lemonade do you need to make 20 litres of this drink?

2   **Reasoning** Jade mixes squash and water in the ratio 3:8.
    Theo mixes squash and water in the ratio 7:20.
    Whose drink is stronger?

    > **Q2 hint** Which has the higher proportion of squash?

3   **Problem-solving** A school announces on its website
    '126 of our students have university places this year – an increase of 17% on 2013.'
    How many students had a university place in 2013?

4   **Exam question**

    The graph shows information about the distances travelled
    by a car for different amounts of petrol used.

    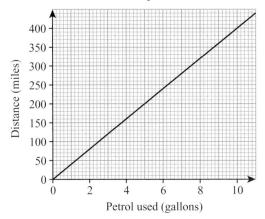

    a   Find the gradient of the straight line.   **(2 marks)**
    b   Write down an interpretation of this gradient.   **(1 mark)**

    June 2014, Q8, 5MB3H/01

    Exam hint
    Many students did not read
    the scale on the horizontal axis
    properly, and so divided by the
    wrong number when finding the
    gradient.

5   **Problem-solving** From 1 November 2014 to 31 January 2015 unemployment fell by
    102 000 to 1.86 million. Work out the percentage fall, to 1 decimal place.

6   Sanjeet recorded these readings in a
    science experiment.
    a   Plot Sanjeet's results on a graph.
    b   Write a formula connecting y and x.

x	0	2	3	6	8	12
y	0	3.2	4.8	9.6	12.8	19.2

7   **Problem-solving** p is inversely proportional to r.
    When $r = 5.6$, $p = 0.75$
    Find r when $p = 1.8$. Give your answer to 3 significant figures.

8   **Problem-solving** Jo and Adam divide £96 between
    them so that Adam has three times as much as Jo.
    How much does Jo get?

9 **Exam question**

An object is travelling at a speed of 2650 metres per second.

How many seconds will the object take to travel a distance of 3.45×10^{10} metres?

Give your answer in standard form, correct to 2 significant figures. **(3 marks)**

November 2010, Q12, 5MB2H/01

Exam hint
A lot of students answering this question calculated the speed correctly but wrote it incorrectly in standard form.

10 **Problem-solving** These two trapezia are mathematically similar.
AG = 3 cm
GD = 6 cm
FE = 9 cm
Work out the length of GF.

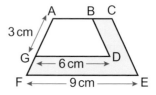

Q10 A common error is to find the length of the whole side, rather than the part the question asks for. Read the question carefully.

11 **Problem-solving** Dan invests £4800 at a compound interest rate of 2.4% per annum for three years. How much interest does he receive in total?

12 **Exam question**

Here are two schemes for investing £2500 for 2 years.

Scheme A

gives 4% **simple** interest each year.

Scheme B

gives 3.9% **compound** interest each year.

Which scheme gives the most total interest over 2 years?

You must show all your working. **(4 marks)**

March 2013, Q13, 5MB3H/01

Exam hint
Make sure you know the difference between simple and compound interest. Many students calculated the wrong type of interest in this question.

13 **Exam question**

The value of a car depreciates by 25% each year.

At the end of 2013 the value of the car was £4800.

Work out the value of the car at the end of 2015. **(3 marks)**

June 2014, Q14, 5MB3H/01

Exam hint
When a value depreciates, it falls. Some students answering this question assumed it had gone up in value from 2013 to 2015.

14 **Problem-solving**
Calculate the density of this 21 kg block of zinc.
Give your answer in g/cm³.

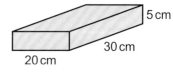

15 **Problem-solving**
Calculate the pressure in N/m² exerted by this container.

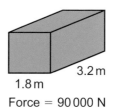

Force = 90 000 N

4 PROBABILITY

What is covered in this section	Foundation Student Book reference
4.1 Venn diagrams • Understand the language of sets and Venn diagrams. • Use Venn diagrams to work out probabilities.	Unit 13
4.2 Frequency trees and tree diagrams • Draw and use frequency trees. • Work out probabilities using tree diagrams.	Unit 13
4.3 Dependent events • Solve probability problems involving events that are not independent.	Unit 13
4.4 Mixed exercise • Consolidate your learning with more practice.	

4.1 Venn diagrams

Objectives

• Understand the language of sets and Venn diagrams.
• Use Venn diagrams to work out probabilities.

Key point 1

Curly brackets { } show a set of values.

∈ means 'is an element of'.

5 ∈ {odd numbers} means '5 is in the set of odd numbers'.

An element is a 'member' of a set. Elements are often numbers, but could be letters, items of clothing or even body parts.

ξ means the universal set – all the elements being considered.

Key point 2

A ∩ B means 'A intersection B'.
This is all the elements that are in A *and* in B.

A ∪ B means 'A union B'.
This is all the elements that are in A *or* B *or* both.

A′ means the elements *not* in A.

Key point 3

You can calculate probabilities from Venn diagrams.

1 Match each diagram to its description.

a   b   c   d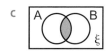

A   A ∩ B   B   A ∪ B   C   B′   D   A ∩ B′

2 A = {positive integers ⩽ 5}
B = {even numbers ⩽ 10}
ξ = {positive integers < 12}

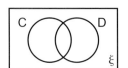

a List the numbers in each set.

Q2a hint A = {1, 2, …}

b Copy and complete the Venn diagram.
c Write 'true' or 'false' for each statement.
 i 6 ∈ A   ii 8 ∈ B   iii 4 ∈ B   iv 2 ∈ A ∩ B   v 7 ∈ A ∪ B   vi 10 ∈ B′

3 C = {positive odd numbers < 10}
D = {prime numbers < 15}
ξ = {positive integers ⩽ 15}

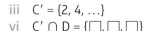

a Copy and complete the Venn diagram.

Q3a hint List the numbers in each set first.

b Copy and complete these sets.
 i C ∩ D = {3, …}   ii C ∪ D = {1, …}   iii C′ = {2, 4, …}
 iv D′ = {1, 4, …}   v C ∩ D′ = {☐, ☐}   vi C′ ∩ D = {☐, ☐, ☐}

Q3b hint Write the numbers in order in your sets.

4 a Draw a Venn diagram to show these sets.
 X = {even numbers, 20 ⩽ x ⩽ 30}
 Y = {multiples of 3, 20 ⩽ y ⩽ 30}
 ξ = {integers, 20 ⩽ n ⩽ 30}

Q4a hint 20 ⩽ x ⩽ 30 means 20 and 30 are included.

b Write these sets.
 i X′   ii X ∩ Y   iii Y′   iv X ∪ Y   v X′ ∩ Y   vi X ∩ Y′

5 People were asked if they had
 • 1 or more brothers (B)
 • 1 or more sisters (S).
The Venn diagram shows the results.
How many people

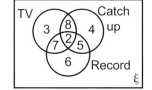

a had brothers but no sisters
b had brothers and sisters
c had no brothers or sisters
d were asked?
e What is the probability that a person picked from this group at random has 1 or more sisters?
f What is the probability that a person picked from this group at random has sisters but no brothers?

6 In a class of 27 students
 • 12 have black hair   • 17 have brown eyes   • 9 have black hair and brown eyes.
a Draw a Venn diagram to show this information.
b What is the probability that a student picked at random from this class
 i has black hair and brown eyes
 ii has black hair but not brown eyes
 iii doesn't have black hair or brown eyes?

7 **Problem-solving** A group of people completed this survey about watching TV.

Tick all that apply.
☐ I watch TV shows as they are broadcast.
☐ I watch TV shows on catch-up websites.
☐ I record TV shows to watch later.
☐ I don't watch any TV shows.

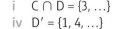

The Venn diagram shows the results.
What is the probability that a person selected at random
a only watches TV shows as they are broadcast

Q7a hint Work out the total number of people surveyed.

b never records TV shows to watch later
c records shows, uses catch-up websites and watches shows as they are broadcast?

8   **Problem-solving** 117 Year 11 students chose the courses they wanted at the
    school prom dinner.
    - 74 chose starter.
    - 102 chose main.
    - 80 chose dessert.
    - 41 chose starter, main and dessert.
    - 26 chose main and dessert.
    - 23 chose starter and main.
    - 8 chose starter and dessert.

Q8 A common error is to not realise that you should draw a Venn diagram.

    Work out the probability that a student picked at random chose main course only.

9   A card is picked at random from this set.

    | 1 | 2 | 3 | 4 | 5 | 6 | 7 | 8 | 9 | 10 |

    The Venn diagram shows two events.
    X = {number < 7}
    Y = {number is even}
    Work out

    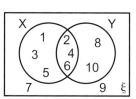

    a   P(X)   b   P(Y)   c   P(X ∩ Y)
    d   P(X ∪ Y)   e   P(X')   f   P(Y')
    g   P(X' ∩ Y)   h   P(X ∪ Y')

Q9a hint P(X) means the probability that a card picked at random is in X.

10 **Reasoning** Maisie asks the 24 students in her class whether they passed their
    science (S) and maths (M) tests.
    - 18 students passed science and maths.
    - 3 students failed both tests.
    - 20 students passed maths.
    a   Draw a Venn diagram to show Maisie's results.
    b   Work out
        i   P(S)   ii   P(S ∩ M)   iii   P(S ∪ M)   iv   P(S' ∩ M)

4.2 Frequency trees and tree diagrams

Objectives

- Draw and use frequency trees.
- Work out probabilities using tree diagrams.

Key point 4

Two events are **independent** when the results of one do not affect the results of the other.

Key point 5

A **frequency tree** shows two or more events and the number of times they occurred.

Key point 6

A **tree diagram** shows two or more events and their probabilities.
For independent events, multiply the probabilities along the branches.

1   For an ordinary, unbiased dice, work out
    a   P(6)   b   P(not 6)   c   P(prime number)
    d   P(not prime)   e   P(square number)   f   P(not square)

2   Tom gets a train to school. Out of 30 school days, the train is late 7 times.
    He missed 3 of the late trains and missed 5 of the trains that were on time.
    Copy and complete the frequency tree.

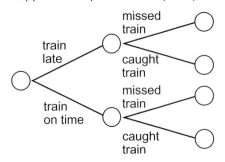

3   40 people complete a survey about pets.
    28 people answer 'No' to the question 'Do you have a dog?'
    Of these 28, 16 answer 'Yes' to the question 'Do you have a cat?'
    4 people who have a dog also have a cat.
    Copy and complete the frequency tree.

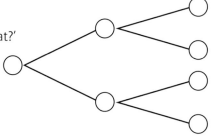

Q3 hint The first pair of branches is for
the answers to 'Do you have a dog?'

4   In a clinical trial 60 people are exposed to a virus.
    After 2 days 44 of them test positive for the virus and the rest test negative.
    15 of the people who tested positive did not develop the virus.
    32 people in total developed the virus.
    a   Copy and complete the frequency tree.

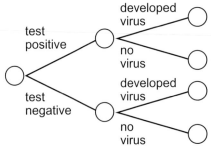

Q4 A common error in frequency
trees is to not check that the two
numbers at the end of each pair of
branches add to make the number
at the beginning of the pair.

    b   How many of the people who tested negative developed the virus?

5   A school has 180 Year 11 students. 96 are girls.
    After Year 11, all the students either stay at school (S), go to college (C), or get an
    apprenticeship (A). 24 boys stay at school. 20 boys get apprenticeships.
    38 girls go to college. 21 girls get apprenticeships.
    a   Copy and complete the frequency tree.

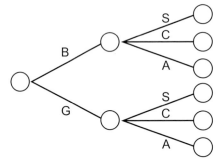

    b   How many students go to college?

6 **Problem-solving** 24 people took their driving test.
15 were under 25 years old. 10 of these passed.
Altogether, 7 people failed.
How many people aged 25 or over passed their test?

Q6 hint Draw a frequency tree.

7 A bag contains red and white counters.
One counter is taken out, the colour is recorded and then it is put back in the bag.
Another counter is then taken out.
The tree diagram shows the probabilities.

1st counter   **2nd counter**

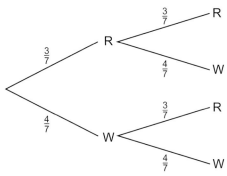

Work out the probability of taking

a   2 reds

b   2 whites

c   1 red and 1 white

d   two counters the same colour.

Q7c hint 1 red and 1 white can be taken in either order – red then white, or white then red.

Q7d hint Two counters the same colour can be either red, red or white, white.

8 At weekends the probability that Jess lies in till 12 is $\frac{2}{3}$.

a   Copy and complete this tree diagram.

Saturday   **Sunday**

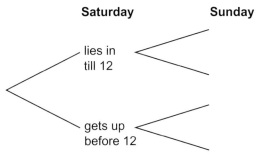

Q8 A common error in tree diagrams is to not check that the probabilities on each pair of branches add to make 1.

b   What is the probability that Jess
   i lies in on both days   ii gets up before 12 on only one day at the weekend?

9 The probability that a spinner stops on red is 0.6. The spinner is spun twice.

a   Copy and complete this tree diagram.

1st spin   **2nd spin**

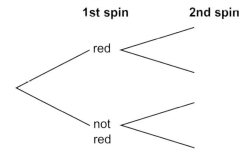

b   Calculate the probability of just one red in two spins.

10 A pack of cards contains 12 picture cards and 40 number cards.
    One card is picked at random and replaced.
    A second card is picked.

    a   Copy and complete the tree diagram.

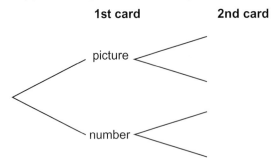

1st card   **2nd card**

picture

number

> **Q10b hint** Use your calculator to
> multiply and add the fractions.

    b   What is the probability of picking one picture card and one number card?

4.3 Dependent events

Objective

• Solve probability problems involving events that are not independent.

Key point 7

If one event depends upon the outcome of another event, the two events are **dependent events**.
For example, removing a red card from a pack of playing cards reduces the chance of choosing
another red card.

Warm up

1   There are 12 coins in a purse. Four are £1 coins and 8 are 10p coins.
    a   Gina takes a coin at random and puts it in a charity box.
        What is the probability that she takes a £1 coin?
    b   If she takes a £1 coin first, what is the probability that the next coin she picks is 10p?

2   Yasmin picks one of these letter cards, puts it to one side, and then picks another.

M A T H E M A T I C S

Copy and complete the frequency tree.

> **Q2 hint** If the first card is a vowel,
> how many vowels are left?

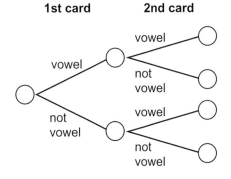

1st card   **2nd card**

vowel

not
vowel

3 A packet of biscuits has 5 chocolate biscuits and 9 plain biscuits.
Grace picks a biscuit at random, eats it, and then picks another.
Use the tree diagram to find the probability that Grace eats

a two chocolate biscuits

b one of each type

c two biscuits of the same type.

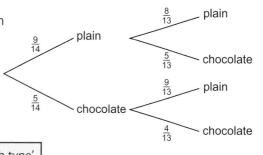

1st biscuit   2nd biscuit

$\frac{9}{14}$ plain
$\frac{8}{13}$ plain
$\frac{5}{13}$ chocolate
$\frac{5}{14}$ chocolate
$\frac{9}{13}$ plain
$\frac{4}{13}$ chocolate

> **Q3b** A common error is to forget that 'one of each type' means plain then chocolate **or** chocolate then plain.

4 Harry has 3 red pens and 4 black pens in his pencil case. He takes out 2 pens.

a Copy and complete the tree diagram.

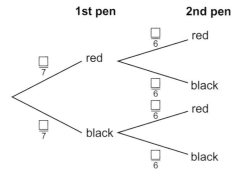

1st pen   **2nd pen**

$\frac{\Box}{7}$ red
$\frac{\Box}{6}$ red
$\frac{\Box}{6}$ black
$\frac{\Box}{7}$ black
$\frac{\Box}{6}$ red
$\frac{\Box}{6}$ black

> **Q4a hint** For this type of question, draw the tree diagram as if he takes out 1 pen, and then another.

b Work out the probability that the pens are different colours.

5 The probability that the maths teacher is in a good mood is 0.7.
When she is in a good mood, the probability that she sets homework is 0.8.
When she is not in a good mood, the probability that she sets homework is 0.9.

a Copy and complete the tree diagram.

b Find the probability that she does not set homework.

0.7 good mood
0.8 sets homework
does not set homework
not good mood
0.9 sets homework
does not set homework

6 Toby has these cards.

21 22 23 24 25 26 27 28 29 30

He picks a card, puts it to one side, and then picks another card.

a Copy and complete the tree diagram.

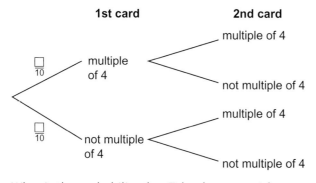

1st card   **2nd card**

$\frac{\Box}{10}$ multiple of 4
multiple of 4
not multiple of 4
$\frac{\Box}{10}$ not multiple of 4
multiple of 4
not multiple of 4

b What is the probability that Toby does not pick any multiples of 4?

7   The probability that Shona goes out on a Friday night is 0.7.
    When she goes out, the probability that she goes to bed after 12 is 0.6.
    When she stays in, the probability that she goes to bed at 12 or earlier is 0.8.
    a   Copy and complete the tree diagram.

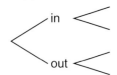

    b   What is the probability that on a Friday night she goes to bed after 12?

8   **Problem-solving** Alix has 5 identical silver earrings
    and 3 identical gold earrings in a box.
    She takes 2 earrings from the box at the same time.
    What is the probability that she picks a matching pair?

> **Q8** A common error is to not realise
> that you should draw a tree diagram
> for this type of question.

9   **Reasoning** Nick takes the train to work.
    The tree diagram shows the probabilities that the train is late, and that he is late,
    on time or early for work.

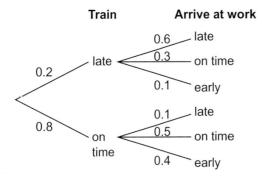

Write 'true' or 'false' for each statement and show working to explain.
    a   When the train is late, he is more likely to be late for work than when the train is on time.
    b   When the train is on time, he is usually at work early.
    c   The probability that he is late for work is more than $\frac{1}{4}$.
    d   Over 10 days, he is unlikely to be early on more than 4 days.

10 Deborah picks 3 counters from a bag containing 10 blue and 7 yellow counters.
    a   Copy and complete the tree diagram.

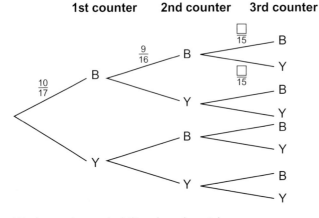

> **Q10** A common error
> is to make mistakes in
> multiplying and adding
> fractions. Remember that it
> is easier to add fractions that
> have not been simplified.

    b   Work out the probability that she picks
        i   3 yellows   ii   2 blues and a yellow.
    c   Which is more likely – 2 blues and a yellow, or 2 yellows and a blue? Explain.

4.4 Mixed exercise

Objective

- Consolidate your learning with more practice.

1 ξ = {positive integers less than 20}
A = {factors of 16}
B = {factors of 18}
C = {first five even numbers}
  a List all the elements of A, B and C.
  b Write down the elements of
    i A ∩ B ⠀⠀ii B ∩ C ⠀⠀iii A ∩ C ⠀⠀iv A ∩ B ∩ C
  c Copy and complete the Venn diagram.

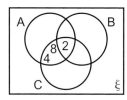

  d List the elements of A ∪ B.

2 ξ = {positive integers ⩽ 24}
D = {factors of 20}
E = {factors of 24}
F = {first seven even numbers}
  a Copy and complete the Venn diagram for these sets.

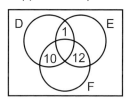

  b List the elements of F ∪ D
  c How many elements are there in E'?

3 **Problem-solving** A group of 50 children choose from two activities – paintballing
and horse riding.
The ratio of boys to girls in the group is 2 : 3.
$\frac{3}{4}$ of the boys choose paintballing.
$\frac{2}{5}$ of the girls choose horse riding.
  a Copy and complete the frequency tree.

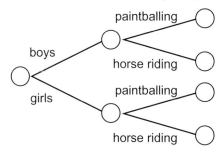

  b How many children choose paintballing?

4

> **Exam question**
>
> There are 40 children in a ski club.
> Each child has one pair of skis.
>
> The skis are twin-tipped skis or downhill skis or slalom skis.
>
> There are 23 boys in the ski club.
> 7 of the boys have twin-tipped skis.
> 8 of the girls have downhill skis.
> 5 of the 9 children with slalom skis are girls.
>
> Work out the number of children with twin-tipped skis.
>
> **(4 marks)**
>
> *February 2013, Q4, 5MB1H/01*

Exam hint
Draw a frequency tree

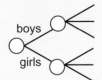

or use a two-way table.

Boys			
Girls			
Total			

5

> **Exam question**
>
> On an activity day students play one sport.
> They play football, hockey or tennis.
> 120 students are on the activity day.
> 30 of the students are boys.
> 12 of the boys and 26 of the girls play hockey.
> 45 of the students play football.
> 35 of the 45 students who play football are girls.
> Work out the number of girls who play tennis.   **(4 marks)**
> *June 2013, Q8, 5MB1II/01*

Exam hint
Students who attempt
to answer this type of
question without drawing
a diagram usually make
more errors.

6

> **Exam question**
>
> 120 children went on a school activities day.
> Some children went bowling.
> Some children went to the cinema.
> The rest of the children went skating.
> 66 of these children were girls.
> 28 of the 66 girls went bowling.
> 36 children went to the cinema.
> 20 of the children who went to the cinema were girls.
> 15 boys went skating.
> Work out the number of children who went bowling. **(4 marks)**
> *November 2010, Q6, 5MB1H/01*

7

> **Exam question**
>
> There are 200 students at a college. Each student studies one
> of art, graphics or textiles.
> Of the 116 female students, 26 study graphics.
> 22 male students study textiles.
> A total of 130 students study art.
> The number of students who study graphics is the same as the
> number of students who study textiles.
> Work out how many male students study art.   **(4 marks)**
> *November 2012, Q6, 5MB1H/01*

Exam hint
Work out how many
students study graphics
or textiles. Then, use
the fact that the same
number study each.

8

> **Exam question**
>
> There are 4 banana smoothies and 3 apple smoothies in a box.
>
> Jenny takes at random 1 smoothie from the box.
> She writes down its flavour, and puts it back in the box.
> Jenny then takes at random a second smoothie from the box.
>
> **a** Complete the probability tree diagram.
>
>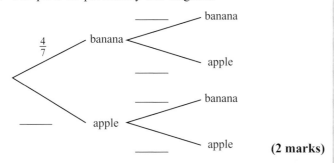
>
> **(2 marks)**
>
> **b** Work out the probability that both smoothies are apple
> flavour. **(2 marks)**
>
> *November 2011, Q13, 5MB1H/01*

Exam hint
Make sure the probabilities on pairs of branches add to 1.

9 **Reasoning** 50 students took English and science tests.
34 students passed both tests.
38 students passed English and 42 students passed science.

a Draw a Venn diagram to show these results.

Zoe says '4% of the students failed both tests.'

b Is she correct? Show working to explain.

10

> **Exam question**
>
> The probability that it will rain on Monday is 0.6.
>
> When it rains on Monday, the probability that it will rain on
> Tuesday is 0.8.
>
> When it does not rain on Monday, the probability that it will
> rain on Tuesday is 0.5.
>
> **a** Complete the probability tree diagram.
>
>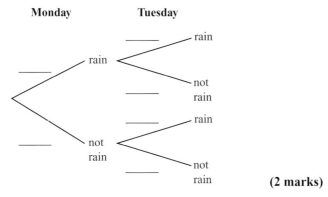
>
> **(2 marks)**
>
> **b** Work out the probability that it will rain on both
> Monday and Tuesday. **(2 marks)**
>
> **c** Work out the probability that it will rain on
> at least one of the two days. **(3 marks)**
>
> *June 2011, Q10, 5MB1H/01*

Exam hint
In this question, some students wrote down the correct calculations, but lost marks by getting the wrong answers. Check your calculations carefully.

Exam hint
In part **c**, rain on at least one day means rain on Monday or Tuesday or both days.
Use P(rain on at least 1 day) =
1 – P(does not rain on either day)

11 ξ is the set of students in a film club.
    A is the set of students who like horror films.
    B is the set of students who like action movies.
    The Venn diagram shows the number of students in each set.
    Work out

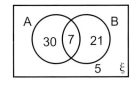

    a P(A ∩ B′)   b P(A′ ∪ B′)

12 **Problem-solving** There are three £1 coins and
    one 10p coin in a box.
    Alina takes two coins at random from the box,
    without replacement.

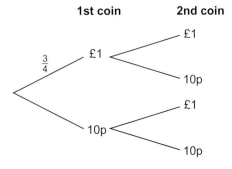

    a Copy and complete the tree diagram.
    b Work out the probability that both coins are £1 coins.
    c Find the probability that both coins left in the
       box are £1 coins.

    ┌─────────────────────────────────┐
    │ Q12c hint If both coins in the box are │
    │ £1, which coins has she taken out? │
    └─────────────────────────────────┘

13 There are 9 counters in a bag.
    • 2 of the counters are yellow.
    • 4 of the counters are red.
    • 3 of the counters are black.
    Tariq takes a counter at random from the bag
    and does not replace it.
    Then he takes a second counter at random
    from the bag.

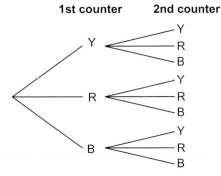

    a Copy and complete the tree diagram.
    b Find the probability that he takes one yellow and
       one black counter.

14 ┌─ **Exam question** ──────────────────────────┐
    │   │
    │ There are 10 pens in a box.   │
    │   │
    │ 4 of the pens are red. 6 of the pens are black. │
    │   │
    │ Josh takes at random a pen from the box.   │
    │ He puts the pen into his bag.   │
    │   │
    │ He then takes at random another pen from the box. │
    │   │
    │ Work out the probability that Josh takes one pen of each │
    │ colour.   **(4 marks)** │
    │   │
    │   *November 2012, Q16, 5MB1H/01*   │
    └──┘

    Exam hint
    Students who did not
    draw a tree diagram for
    this question were more
    likely to get the answer
    wrong.

15 ┌─ **Exam question** ──────────────────────────┐
    │   │
    │ Here are some cards. Each card has a letter on it. │
    │   │

    │ Rachel takes at random two of these cards. │
    │   │
    │ Work out the probability that there are different letters on the │
    │ two cards.   **(4 marks)** │
    │   │
    │   *June 2012, Q14, 5MB1H/01*   │
    └──┘

    Exam hint
    Many students answering
    this question assumed
    incorrectly that once
    a card was taken it
    was replaced. But
    this question is about
    conditional probability
    – the first card is not
    replaced before the next
    one is taken.

5 GEOMETRY AND MEASURES

What is covered in this section	Foundation Student Book reference
5.1 Problem-solving • Solve angle problems.	Unit 6
5.2 Transformations • Transform shapes using one or more transformations. • Describe transformations of shapes on a grid.	Unit 10
5.3 Vectors • Write, add and subtract vectors. • Solve vector problems.	Unit 19
5.4 Right-angled triangles • Use Pythagoras theorem and trigonometry to solve problems in right-angled triangles.	Unit 12
5.5 Constructions, bearings and loci • Construct scale diagrams and loci to solve problems.	Unit 15
5.6 Perimeter, area and volume • Find perimeter, area and volume of 2-D shapes and 3-D solids.	Units 8, 17
5.7 Congruence • Show that two triangles are congruent. • Find missing sides and angles in congruent shapes.	Unit 19
5.8 Mixed exercise • Consolidate your learning with more practice.	

5.1 Problem-solving

Objective

• Solve angle problems.

Key point 1

In parallel lines

• **alternate** angles are equal

• **corresponding** angles are equal.

Key point 2

The sum of the **exterior angles** of a regular polygon is always **360°**.

The sum of the **interior angles** of a polygon with n sides is $(n - 2) \times 180°$.

In a **regular polygon**, all the angles are the same size, so exterior angle = $\dfrac{360°}{\text{number of sides}}$

1 Work out the sizes of the angles marked with letters.

a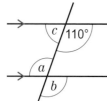

b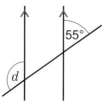

c

2 a Calculate the size of one exterior angle of a regular nonagon (9 sides).

   b What size is its interior angle?

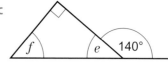
Q2 hint   exterior   interior

3 A regular octagon is divided into 8 isosceles triangles.
Work out the size of

   a angle d

   b angle e

   c angle f

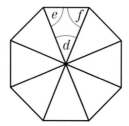

Q3a hint Angles at a point add to ☐

4 Reasoning **a** Copy this regular hexagon. Divide it into 6 isosceles triangles.

   b Find the sizes of all the angles in one of the 6 triangles.

   c What type of triangles are they? Explain.

5 Work out the sizes of the exterior angles in this polygon.

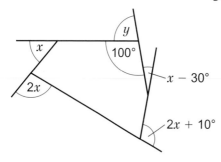

Q5 hint The question does not say it is regular, so don't assume the exterior angles are equal.

Q5 A common error is to find the value of x, but then not use the value to find the size of each angle.

6 Reasoning Melissa is asked to work out the missing angles in this shape.

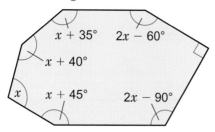

Here is Melissa's answer. It is incorrect.

7 sides, sum of interior angles = $(7 - 2) \times 180° = 900°$

$x + 35 + 2x - 60 + x + 40 + x + 45 + 2x - 90 + x = 900°$

$8x - 30 = 900°$

$8x = 930°$

$x = \dfrac{930}{8} = 116.25°$

   a Explain her mistakes.

   b Work out the correct answer.

7   ABCDEF is a regular hexagon. ABDE is a rectangle.

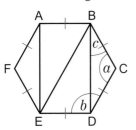

Q7b hint An isosceles trapezium has equal sides and equal angles as shown.

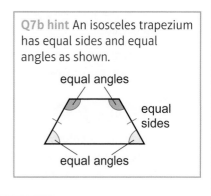

equal angles

equal sides

equal angles

  a   Work out the sizes of angles a, b and c. Give reasons.
  b   Show that BCDE is an isosceles trapezium.
  c   Find the size of angle BED. Give reasons.

8   **Reasoning** Find the sizes of the angles labelled with letters in this regular octagon. Give reasons for each step.

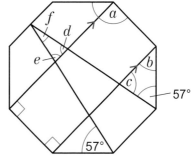

9   **Reasoning** The diagram shows a triangle drawn from the centre of a regular polygon. The exterior angle of the polygon is 30°.
  a   Work out the size of angle y. What do you notice?

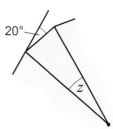

  b   Angle z is in a triangle drawn from the centre of a regular polygon with exterior angle 20°. Predict the value of z. Show working to check your prediction.

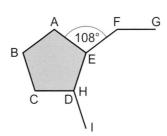

10 **Problem-solving** One side of a regular pentagon forms the side of a regular polygon with n sides. Angle AEF = 108° Work out the value of n.

11 **Problem-solving** The diagram shows the interior and exterior angles of a regular polygon. How many sides does the polygon have?

Q11 hint Find x first.

12 **Problem-solving** In a regular polygon, the interior angle is 3 times the size of the exterior angle. Name the polygon.

5.2 Transformations

Objectives

- Transform shapes using one or more transformations.
- Describe transformations of shapes on a grid.

Key point 3

You can use a **column vector** to describe a **translation**.

The top number describes the movement to the left or right, and the bottom number describes the movement up or down. For example:

$\begin{pmatrix} 3 \\ 2 \end{pmatrix}$ means 3 right, 2 up $\qquad$ $\begin{pmatrix} -4 \\ -5 \end{pmatrix}$ means 4 left, 5 down

Key point 4

To describe a **reflection** on a coordinate grid you need to give the equation of the **mirror line**.
To describe a **rotation** you need to give the **angle**, the **direction of turn** and the **centre of rotation**.
To describe an **enlargement** you need to give the **scale factor** and the **centre of enlargement**.

1   a   Copy shape A onto a coordinate grid.
    b   Reflect shape A in the line $y = -1$.
       Label the image B.
    c   Rotate shape A 90° anticlockwise about (0, 0).
       Label the image C.
    d   Enlarge shape A by scale factor 2, centre (0, 2).
       Label the image D.
    e   Translate shape A by $\begin{pmatrix} -5 \\ -8 \end{pmatrix}$.
       Label the image E.

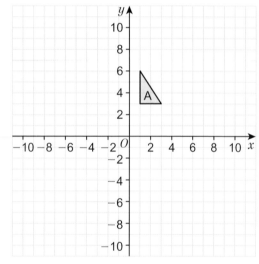

2   a   Copy the diagram.
    b   Translate shape B by $\begin{pmatrix} 6 \\ 1 \end{pmatrix}$.
       Label the image C.
    c   Translate shape C by $\begin{pmatrix} 2 \\ -5 \end{pmatrix}$.
       Label the image D.
    d   Translate shape D by $\begin{pmatrix} -8 \\ 1 \end{pmatrix}$.
       Label the image E.
    e   Write the column vector that maps
       i   E to C   ii   B to D   iii   B to E
       iv   D to C   v   E to D   vi   E to B

> **Q2e** A common error is to write vectors like coordinates, for example writing (6, 2) instead of $\begin{pmatrix} 6 \\ 2 \end{pmatrix}$.

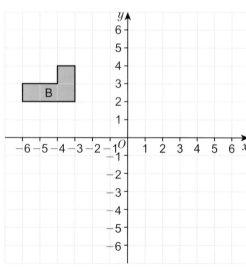

3 a Copy the coordinate grid and shapes F and G.
   b Reflect shape F in the line $y = -x$.

   > **Q3b hint** Draw in the line before
   > you do the reflection.

   c Reflect shape G in the line $y = x$.
   d Describe the reflection that takes
      i   shape H to shape J
      ii shape K to shape L.
   e Describe the rotation that takes shape J
      to shape H.

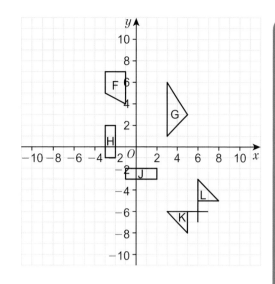

4 a Copy the diagram.

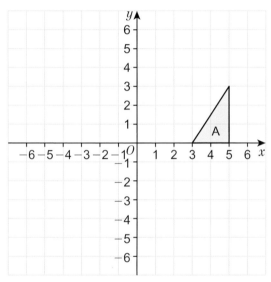

   b Rotate shape A 180° about (1, 2). Label the image B.
   c Rotate shape B 90° anticlockwise about (0, −2). Label the image C.
   d Describe the rotation that maps shape A onto shape C.

5 a Copy the diagram and shapes C and D.
   b Enlarge shape C by scale factor $\frac{1}{3}$, centre (0, 3).
   c Enlarge shape D by scale factor $\frac{1}{4}$, centre (4, 5).
   d Describe the enlargement that takes
      i   shape E to shape F
      ii shape F to shape E.

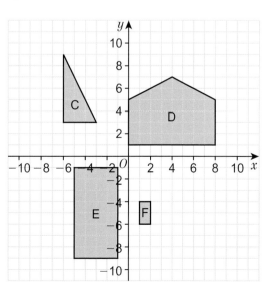

6 a Describe fully the single transformation that maps triangle M onto triangle N.

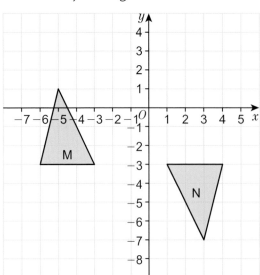

Q6b, c hint The coordinates do not have to be whole numbers.

b Copy the diagram. Reflect triangle M in the y-axis and label the image P.

c Describe fully the single transformation that maps P onto N.

7 **Problem-solving**

a There are two possible single transformations that map triangle Q onto triangle R. Describe them both.

b There are two possible single transformations that map triangle Q onto triangle S. Describe them both.

c Copy the diagram. Rotate triangle S 90° clockwise about (6, −3). Label the image T.

d Describe the single transformation that maps
  i triangle T onto triangle R
  ii triangle Q onto triangle T.

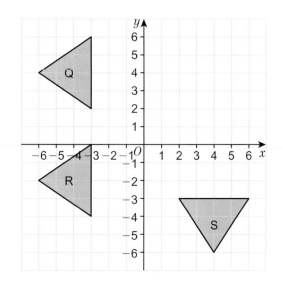

8 **Problem-solving**

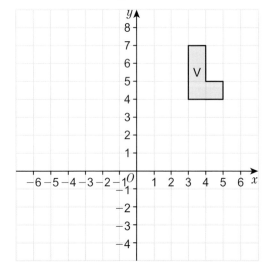

a Copy the diagram. Reflect shape V in the line $x = 1$. Label the image U.

b Reflect shape U in the line $y = 3$. Label the image W.

c Describe the single transformation that takes shape V to shape W.

d Shape U is reflected in the line $x = k$ to give shape X.

The translation $\begin{pmatrix} -8 \\ 0 \end{pmatrix}$ takes shape V to shape X.

Find the value of k.

Q8d hint Draw the image of V after the translation.

9 a Copy the diagram.

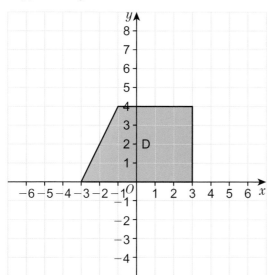

b Enlarge shape D by scale factor $\frac{1}{2}$, centre (1, 0). Label the image E.

c Translate shape E by $\begin{pmatrix} -1 \\ -2 \end{pmatrix}$. Label the image F.

d Describe the single transformation that takes shape D to shape F.

10 **Reasoning** Marc says, 'In transformations, the image is always congruent to the original shape'. Draw or describe an example to show why Marc is wrong.

11 **Problem-solving**

a Draw the triangle with vertices at (4, 6), (6, 6) and (5, 3) on a coordinate grid.

> **Q11 communication hint** A **vertex** is a corner. The plural of vertex is **vertices**.

b Describe a transformation that maps the triangle to the third quadrant:
   i   the same way up
   ii facing in a different direction.

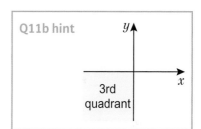

Q11b hint

3rd quadrant

5.3 Vectors

Objectives

• Write, add and subtract vectors.
• Solve vector problems.

Key point 5

To add two column vectors, add the top numbers and add the bottom numbers.

$$\begin{pmatrix} 2 \\ 3 \end{pmatrix} + \begin{pmatrix} 4 \\ 1 \end{pmatrix} = \begin{pmatrix} 6 \\ 4 \end{pmatrix}$$

To subtract two column vectors, subtract the top numbers and subtract the bottom numbers.

$$\begin{pmatrix} 5 \\ 7 \end{pmatrix} - \begin{pmatrix} 3 \\ 4 \end{pmatrix} = \begin{pmatrix} 2 \\ 3 \end{pmatrix}$$

Key point 6

−**b** is the negative of **b** and points in the opposite direction.

If $\mathbf{b} = \begin{pmatrix} -2 \\ 3 \end{pmatrix}$, then $-\mathbf{b} = \begin{pmatrix} 2 \\ -3 \end{pmatrix}$

a − **b** is the same as **a** + (−**b**).

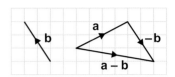

Key point 7

You can multiply a vector by a number.

For example, if $\mathbf{a} = \begin{pmatrix} 4 \\ 3 \end{pmatrix}$ then $2\mathbf{a} = \begin{pmatrix} 8 \\ 6 \end{pmatrix}$ and $-3\mathbf{a} = \begin{pmatrix} -12 \\ -9 \end{pmatrix}$.

1   Write as column vectors.

a $\overrightarrow{AB}$ b $\overrightarrow{CD}$ c $\overrightarrow{EF}$ d $\overrightarrow{GH}$

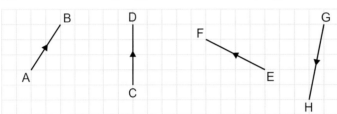

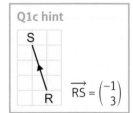

Q1c hint

$\overrightarrow{RS} = \begin{pmatrix} -1 \\ 3 \end{pmatrix}$

2   Add these column vectors.

a $\begin{pmatrix} 5 \\ 3 \end{pmatrix} + \begin{pmatrix} 0 \\ 2 \end{pmatrix}$

b $\begin{pmatrix} -3 \\ 1 \end{pmatrix} + \begin{pmatrix} 2 \\ 3 \end{pmatrix}$

c $\begin{pmatrix} 5 \\ 4 \end{pmatrix} + \begin{pmatrix} -2 \\ 1 \end{pmatrix}$

d $\begin{pmatrix} -3 \\ 7 \end{pmatrix} + \begin{pmatrix} -2 \\ -5 \end{pmatrix}$

Q2 A common error is to make mistakes adding negative numbers.

3   a Write as column vectors.

   i $\overrightarrow{XY}$

   ii $\overrightarrow{YZ}$

   iii $\overrightarrow{XZ}$

b Use the column vectors from part **a** to show that $\overrightarrow{XY} + \overrightarrow{YZ} = \overrightarrow{XZ}$.

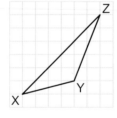

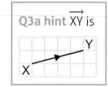

Q3a hint $\overrightarrow{XY}$ is

4   Draw these vectors on squared paper.

a $\mathbf{a} = \begin{pmatrix} 3 \\ 5 \end{pmatrix}$ b $\mathbf{b} = \begin{pmatrix} -1 \\ 4 \end{pmatrix}$ c $\overrightarrow{ST} = \begin{pmatrix} -2 \\ -7 \end{pmatrix}$ d $\overrightarrow{PQ} = \begin{pmatrix} 4 \\ -3 \end{pmatrix}$

5   Find the missing vectors.

a $\begin{pmatrix} 2 \\ 1 \end{pmatrix} + \begin{pmatrix} \ \\ \ \end{pmatrix} = \begin{pmatrix} 5 \\ 3 \end{pmatrix}$

b $\begin{pmatrix} \ \\ \ \end{pmatrix} + \begin{pmatrix} -5 \\ 4 \end{pmatrix} = \begin{pmatrix} -2 \\ 3 \end{pmatrix}$

c $\begin{pmatrix} 7 \\ 5 \end{pmatrix} - \begin{pmatrix} \ \\ \ \end{pmatrix} = \begin{pmatrix} 5 \\ 4 \end{pmatrix}$

d $\begin{pmatrix} \ \\ \ \end{pmatrix} - \begin{pmatrix} 4 \\ 5 \end{pmatrix} = \begin{pmatrix} 3 \\ -3 \end{pmatrix}$

e $\begin{pmatrix} -2 \\ 3 \end{pmatrix} + \begin{pmatrix} \ \\ \ \end{pmatrix} = \begin{pmatrix} -5 \\ 2 \end{pmatrix}$

f $\begin{pmatrix} \ \\ \ \end{pmatrix} - \begin{pmatrix} -2 \\ 0 \end{pmatrix} = \begin{pmatrix} 6 \\ 6 \end{pmatrix}$

6   **Problem-solving**

$\mathbf{a} = \begin{pmatrix} 3 \\ 3 \end{pmatrix}$

$\mathbf{a} + \mathbf{b} = \begin{pmatrix} 4 \\ 8 \end{pmatrix}$

Find **b**.

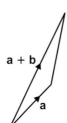

7 Point A has coordinates (1, 2). Point B has coordinates (4, 6).
   a Plot points A and B on a coordinate grid.
   b The vector $\overrightarrow{BC}$ is $\begin{pmatrix} 4 \\ -2 \end{pmatrix}$. Draw this on your grid.
   c Find the coordinates of C.
   d Find the vector $\overrightarrow{AC}$.

Q7 A common error is to confuse coordinates (3, 2) and vectors $\begin{pmatrix} 3 \\ 2 \end{pmatrix}$.

8 **Problem-solving** Point X has coordinates (0, 1).
   Point Y has coordinates (−2, 4).
   $\overrightarrow{YZ} = \begin{pmatrix} 4 \\ 3 \end{pmatrix}$
   Find
   a the coordinates of Z
   b $\overrightarrow{XZ}$

9 $\mathbf{p} = \begin{pmatrix} 3 \\ 1 \end{pmatrix}$ $\mathbf{q} = \begin{pmatrix} -6 \\ 5 \end{pmatrix}$ $\mathbf{r} = \begin{pmatrix} -2 \\ -5 \end{pmatrix}$ $\mathbf{s} = \begin{pmatrix} 4 \\ -1 \end{pmatrix}$
   Find the resultant of
   a **p** and **q**   b **q** and **r**   c **p** and **s**   d **r** and **p**

Q9 communication hint
Resultant of **a** and **b** is **a** + **b**

10 The vectors **d**, **e** and **f** are shown on the grid.
    Write as column vectors
    a **d**   b 2**d**   c −**e**
    d **d** + **e**   e **f** − **e**   f 3**e**
    g −2**f**   h **d** − 2**f**

    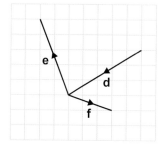

    Q10b hint 2**d** = 2 × **d** = 2 × $\begin{pmatrix} \\ \end{pmatrix}$

11 **Problem-solving** ABCD is a quadrilateral.

    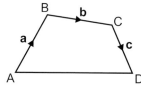

    Write in terms of **a**, **b** and **c**
    a $\overrightarrow{AC}$ b $\overrightarrow{BD}$ c $\overrightarrow{AD}$ d $\overrightarrow{CA}$

Q11a hint

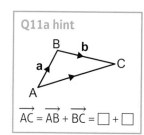

$\overrightarrow{AC} = \overrightarrow{AB} + \overrightarrow{BC} = \square + \square$

12 **Problem-solving** PQRS is a quadrilateral.

    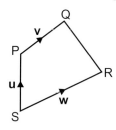

    a Write in terms of **u**, **v** and **w**
       i $\overrightarrow{PS}$ ii $\overrightarrow{SQ}$ iii $\overrightarrow{QP}$ iv $\overrightarrow{PR}$
    b Show that $\overrightarrow{QR} = -\overrightarrow{RQ}$

Q12b hint Find $\overrightarrow{QR}$ and $\overrightarrow{RQ}$ in terms of **u**, **v** and **w**.

5.4 Right-angled triangles

Objective

- Use Pythagoras' theorem and trigonometry to solve problems in right-angled triangles.

Key point 8

Pythagoras' theorem shows the relationship between the lengths of the three sides of a right-angled triangle.

$c^2 = a^2 + b^2$

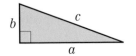

Key point 9

You need to know these ratios and be able to choose the one you need to solve a problem.

$\sin \theta = \dfrac{\text{opposite}}{\text{hypotenuse}}$ $\qquad$ $\cos \theta = \dfrac{\text{adjacent}}{\text{hypotenuse}}$ $\qquad$ $\tan \theta = \dfrac{\text{opposite}}{\text{adjacent}}$

Key point 10

The angle of **elevation** is the angle measured upwards from the horizontal.

The angle of **depression** is the angle measured downwards from the horizontal.

1 Solve each equation to find the unknown. Give your answers to 1 d.p.

 a $c^2 = 3^2 + 5^2$ $\qquad$ b $8^2 = 2^2 + b^2$ $\qquad$ c $\sin 50° = \dfrac{x}{6}$

 d $\cos 20° = \dfrac{5}{x}$ $\qquad$ e $\tan \theta = \dfrac{5}{3}$ $\qquad$ f $\sin \theta = \dfrac{2}{7}$

2 Find the unknown lengths in these right-angled triangles. Give your answers to 2 d.p.

 a
 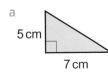
 5 cm
 7 cm

 b
 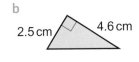
 2.5 cm $\qquad$ 4.6 cm

 Q2 A common error is to not realise which side is the hypotenuse.

3 Find the unknown lengths in these triangles.
 Give your answers in surd form.

 Q3 hint Give the answer $\sqrt{\square}$ or $\square\sqrt{\square}$ from your calculator. Don't convert to a decimal.

 a
 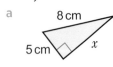
 8 cm
 5 cm $\quad$ x

 b
 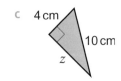
 17 cm $\quad$ 13 cm
 y

 c
 4 cm
 10 cm
 z

4 **Reasoning** When the line segments AB, CD and EF are drawn on a centimetre grid, which is the longest?
 Show working to explain.

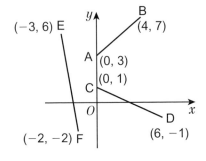

$(-3, 6)$ E $\qquad$ B $(4, 7)$

A $(0, 3)$

C $(0, 1)$

$(-2, -2)$ F

D $(6, -1)$

Diagram NOT accurately drawn

5 **Problem-solving** Find the missing length in each triangle.
   All lengths are in cm.
   Give your answers to 1 d.p.

Q5 A common error is to use the wrong ratio. Make sure you label opp, hyp and adj carefully.

a

b

c

6 **Problem-solving** Find the missing angle in each triangle.
   All lengths are in cm.
   Give your answers to the nearest degree.

a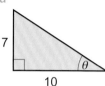

b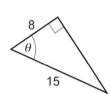

c

d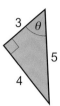

7 **Problem-solving** Find the missing lengths, to 1 d.p.

Q7 hint Use the units in the question.

a

b

c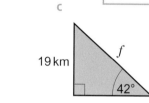

8 **Problem-solving** The angle of elevation to the top of a mobile phone mast is 72° from a point at ground level 5 m from the mast. Find the height of the mast, to the nearest cm.

Q8 hint Draw a diagram.

9 **Problem-solving** From the top of a cliff 60 m high, the angle of depression to a buoy is 40°. How far is the buoy from the bottom of the cliff?

10 a Copy these triangle sketches.
      Find the missing lengths in surd form.
      Write them on your diagrams.
   b Use your triangles from part **a** to help you find:
      i   sin 30°   ii   sin 45°   iii   sin 60°
      iv   cos 30°   v   cos 45°   vi   cos 60°
      vii tan 30°   viii tan 45°   ix   tan 60°
   c Which of these   i   have value 0   ii   have value 1?

| cos 0° | cos 90° | sin 90° | sin 0° | tan 0° |

11 **Reasoning** Use your triangles from **Q10** to show that
   a cos 60° = sin 30°
   b cos 45° = sin 45°

5.5 Constructions, loci and bearings

Objective

• Construct scale diagrams and loci to solve problems.

Key point 11

A **perpendicular bisector** cuts a line in half at right angles

Points **equidistant** from two points lie on the perpendicular bisector of the line joining the two points.
The **shortest path** from a point to a line is perpendicular to the line.
An **angle bisector** cuts an angle exactly in half.

70°

Points equidistant from two lines lie on the angle bisector.

Key point 12

A **locus** is a set of points that obey a given rule. This produces a path followed by the points.
The plural of locus is **loci**.

Key point 13

A **bearing** is an angle measured in degrees clockwise from north.
A bearing is always written using three digits.
This bearing is 025°.

N

25°

---

1 a Draw a line 10 cm long. Construct its perpendicular bisector.
  b Use a protractor to draw an angle of 70°. Construct its angle bisector.

2 a Draw a line and mark a point C on it.
    Mark a point P somewhere above your line.
    Construct the perpendicular from P to the line.
  b Construct the perpendicular at C.

> **Q2a hint**
>
> • P
>
> ———•———
>
> C

3 **Reasoning** The scale diagram shows the position of a walker in a field. There is a bull in the field.
  a Trace the diagram and construct the shortest path to each fence.
  b The walker wants to get out of the field as quickly as possible. Should she walk south or north-east? Explain.

> **Q3** A common error is to rub out construction arcs. Remember, arcs get marks.

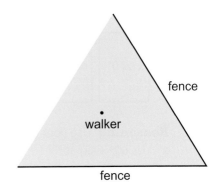

fence

• walker

fence

4 **Problem-solving** The diagram shows the positions of a ship and two walls of a harbour.

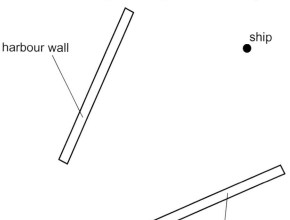

Q4 hint Extend the harbour walls on your diagram until they meet.

The ship leaves the harbour on a course equidistant from the two walls.
Trace the diagram and construct its course.

5 This equilateral triangle is placed on a horizontal surface and rotated three times clockwise about the bottom right-hand vertex.
Sketch the locus of vertex A.

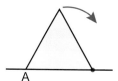

Q5 hint Start by sketching the triangle and vertex A in each new position.

6 **Problem-solving** Three streetlights are placed 15 m apart along a path.
Each light casts light 10 m in all directions.
By constructing a scale diagram, show that an area at least 5 m either side of the line of lights is lit by the lights. Use a scale of 1 cm represents 5 m.

7 **Problem-solving** A plane takes off from an airport and flies east for 12 km and then on a bearing of 140° for 20 km.
   a How far is it from the airport?
   b What bearing does it need to fly on to return to the airport?

8 Accurately construct an angle of 135° using a straight edge and compasses.

Q8 hint First construct a perpendicular bisector. Then bisect one of the 90° angles.

9 **Problem-solving** The scale diagram shows the positions of two towns, Lowtown and Woodville.

Woodville
×

Lowtown

Scale: 1 cm represents 5 km

A company wants to build a warehouse less than 25 km from Lowtown and less than 30 km from Woodville.
The warehouse must be closer to Lowtown than to Woodville.
Trace the diagram and shade the region where the company can build the warehouse.

5.6 Perimeter, area and volume

Objective

• Find the perimeter, area and volume of 2D shapes and 3D solids.

Key point 14

For a **sector** of a circle with an angle of $x°$ and radius r:

arc length $= \dfrac{x}{360} \times 2\pi r$

area of sector $= \dfrac{x}{360} \times \pi r^2$

Key point 15

The volume of a **pyramid** $= \frac{1}{3} \times$ area of base $\times$ vertical height

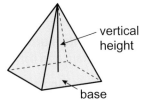

vertical height

base

Key point 16

The volume of a **cone** $= \frac{1}{3} \times$ area of base $\times$ vertical height
The **slant height** l of a cone is the length of the sloping side.
The area of the curved surface of a cone $= \pi \times$ base radius $\times$ slant height $= \pi r l$

Key point 17

The volume of a **sphere** $= \frac{4}{3}\pi r^3$
The surface area of a sphere $= 4\pi r^2$

---

1 Calculate

  a the area

  b the circumference

  of this circle.

6 cm

> **Q1 hint** Give your answers to 1 d.p.

2 Calculate the volume of these solids, to 3 s.f.

  a

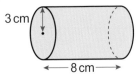
3 cm
8 cm

  b

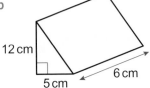
12 cm
5 cm
6 cm

> **Q2a hint**
> Volume of a cylinder $= \pi r^2 h$

3 This prism has volume 25 cm³.
Find

  a the area of its cross-section, to 2 d.p.

  b the height of the prism, to 1 d.p.

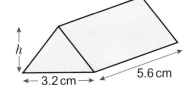
h
3.2 cm
5.6 cm

Warm up

4 **Problem-solving**
Work out the shaded area, to 3 s.f.

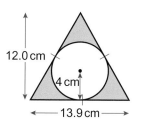

5 **Problem-solving** Penny wants to dig up her lawn to make a flowerbed in the shape of a quarter circle.
Penny says, 'The flowerbed will use up one third of the lawn space'.
Is Penny correct?

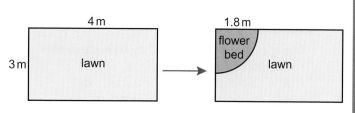

6 **Problem-solving** These two drinks cans both hold 330 ml.

A   B

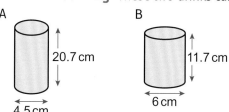

They are made from aluminium 0.5 mm thick.
Which uses less aluminium?

Q6 A common error is to forget to calculate the areas of the top and bottom faces.

7 **Problem-solving** Which sector has the largest area?

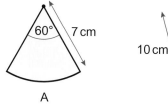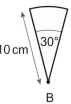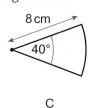

A   B   C

8 **Problem-solving** Polystyrene floats are made in the shape of a cylinder with a hemisphere at each end.

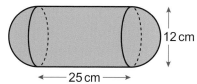

Q8 hint Work out the radius of the hemisphere and the cylinder.

Work out the volume of polystyrene needed to make one float, to the nearest cm³.

9 **Problem-solving** A plastic bead is a sphere of radius 45 mm, with a cylindrical hole through its centre.
The diameter of the hole is 8 mm.
Calculate the volume of plastic in the bead, to the nearest mm³.

10 A yurt is a type of tent made from felt.
It is in the shape of a cylinder, topped with a cone.
Calculate the amount of felt needed for this yurt.
Give your answer to 3 s.f.

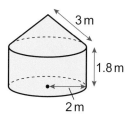

Q10 hint The yurt does not have a floor.

5.7 Congruence

Objectives

- Show that two triangles are congruent.
- Find missing sides and angles in congruent shapes.

Key point 18

Triangles are **congruent** if they have equivalent:
- SSS (all three sides)
- ASA (two angles and the included side)
- SAS (two sides and the included angle
- RHS (right angle, hypotenuse, side)

Triangles where all angles are the same (AAA) are **similar**, but might not be congruent.

Key point 19

When two shapes are congruent, one can be rotated or reflected to fit exactly on the other.

Warm up

1   In each pair, the two triangles are congruent.
    Give reasons to explain why.

> **Q1 hint** Label the side(s) you know with S, any right angles with R, and any other angle(s) you know with A. Do you have SSS, SAS, ASA or RHS?

a
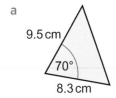

b
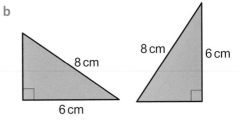

c
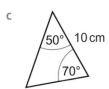

d
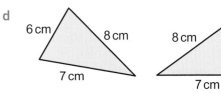

2   **Reasoning** Decide if the triangles in each pair
    are congruent. If they are, give the reason.

> **Q2 hint** Label sides (S) and angles (R or A).

a

b
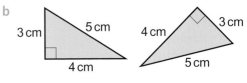

c
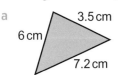
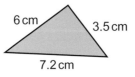

d
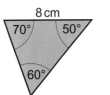
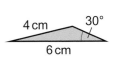

3 **Reasoning** Show that the triangles in each pair are congruent. Give the reason in each case.

Q3 hint Find missing sides and angles.

a
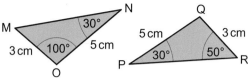

b
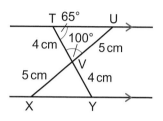

c
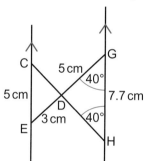

4 **Reasoning** Are triangles TUV and VXY congruent? Explain.
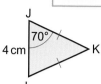

5 **Reasoning** Are triangles CDE and GDH congruent? Explain.

Q5 A common error is to assume that two equal lengths are corresponding sides.

6 **Problem-solving** Find lengths
a x and y
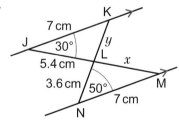

b a and b.
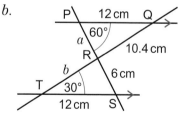

Q6a hint First show the triangles are congruent.

7 **Reasoning** ABCD is a kite.
a Sketch ABCD and mark equal lengths with the same number of dashes.
b Draw in the diagonal BD. Mark equal angles with the same number of arcs.
c Show that BC divides the kite into two congruent triangles.

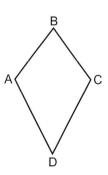

Objective

• Consolidate your learning with more practice.

1   **Problem-solving** The diagram shows a tent.
    Tent fabric costs £5.60 per square metre.
    You can only buy the fabric in whole numbers
    of square metres.
    Work out the cost of the fabric to make this tent,
    to the nearest £.

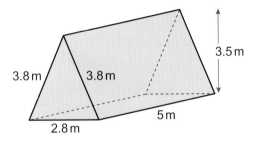

| Q1 hint The tent does not have a floor. |

2   The diagram shows a playground in the shape of a rectangle.

| Q2 A common error is to not draw complete arcs that meet the rectangle. |

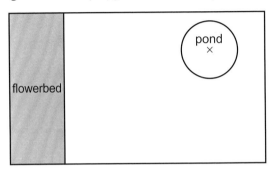

Scale: 1 cm represents 4 m

Marta is going to plant a tree in the playground.
The tree must be
• more than 6 metres from the flowerbed
• **and** more than 8 metres from the centre of the pond.
Copy the diagram, and shade the region where Marta can plant the tree.

3   **Exam question**

The diagram shows a parallelogram.

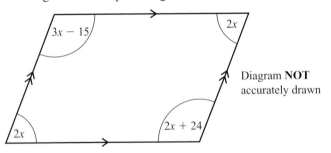

Diagram **NOT** accurately drawn

The sizes of the angles, in degrees, are
    $2x$
    $3x - 15$
    $2x$
    $2x + 24$
Work out the value of x.   **(3 marks)**

June 2012, Q11, 1MA0/1H

Exam hint
Most students did not check their answer by substituting back into the expressions for the angles and checking they added to the correct total. If they had, some may have realised they had made an error in their calculations.

Exam question

CDEF is a straight line.

AB is parallel to CF.

DE = AE

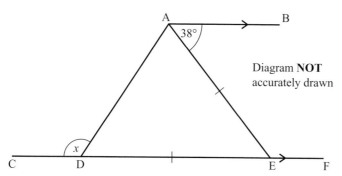

Diagram **NOT** accurately drawn

Work out the size of the angle marked x.

You must give reasons for your answer. **(4 marks)**

February 2013, Q10, 1MA0/1H

> **Q4** A common error is to not write the full reason in angle problems. For example 'angles in a triangle' is not enough. You need to write 'angles in a triangle sum to 180°'.

5 **Problem-solving** Copy this scale drawing of a reservoir. Kim sails from A on a bearing of 060° for 120 metres.

a Use a ruler and protractor to draw her course accurately on the diagram.

b Her boat begins to leak. Construct the course that is the shortest possible distance to a bank of the reservoir.

c Write the bearing and distance for the course you constructed in part **b**.

Scale: 1 cm represents 20 m

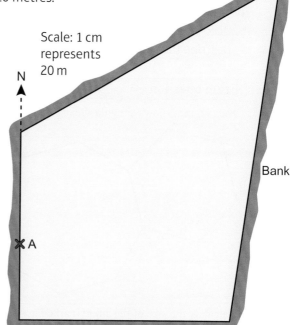

Bank

6 **Problem-solving** The diagram shows a design for a slide. Work out the sizes of the angles marked with letters.

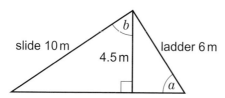

7

Exam question

ABCDE is a regular pentagon.

BCF and EDF are straight lines.

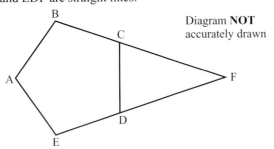

Diagram **NOT** accurately drawn

Work out the size of angle CFD.

You must show how you got your answer. **(3 marks)**

June 2014, Q11, 1MA0/1H

Q7 A common error is to confuse the exterior and interior angles. Which ones add up to 360°?

8 **Reasoning** Shape A is reflected in the line $x = 3$.

The image is reflected in the y-axis to give image B.

a Describe the single transformation that takes A to B.

Shape B is reflected in the line $y = 1$.

The image is reflected in the line $x = 1$ to give image C.

b Describe the single transformation that takes B to C.

c Describe the single transformation that maps A to C.

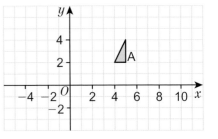

9

Exam question

The diagram shows a regular hexagon and a regular octagon.

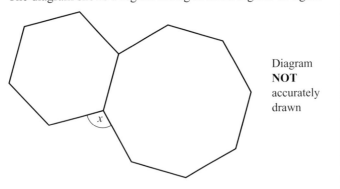

Diagram **NOT** accurately drawn

Calculate the size of the angle marked x.

You must show all your working. **(4 marks)**

June 2012, Q13, 1MA0/1H

Exam hint

You can always go back to basics and divide a polygon into triangles to work out the sum of its interior angles.

10 **Problem-solving** A is the point (2, 1) drawn on a centimetre grid.

$$\overrightarrow{AB} = \begin{pmatrix} 0 \\ 4 \end{pmatrix} \text{ and } \overrightarrow{AC} = \begin{pmatrix} 7 \\ 0 \end{pmatrix}$$

a Find the coordinates of C.

b Calculate the length of BC to the nearest cm.

c Write the column vector $\overrightarrow{BC}$.

Q10a hint Draw A and the vectors on a grid.

11 Exam question

The diagram shows a square and 4 regular pentagons.

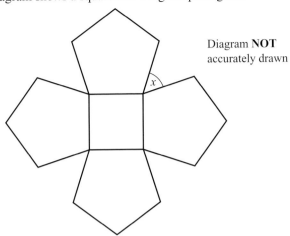

Diagram **NOT** accurately drawn

Work out the size of the angle marked x. **(3 marks)**

March 2013, Q13, 1MA0/1H

12 **Problem-solving** The minute hand of a clock is 24 cm long.

a Calculate the distance travelled by the tip of the minute hand in 10 minutes.

b Calculate the speed of the tip of the minute hand, in centimetres per minute.

Q12 hint What angle does the hand travel through in 10 minutes?

13 Exam question

The diagram shows the positions of three wind turbines A, B and C.

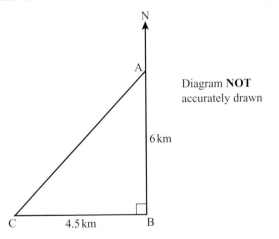

Diagram **NOT** accurately drawn

Exam hint
The question asks you to calculate the bearing, so measuring the angle gains no marks.

A is 6 km due north of turbine B.

C is 4.5 km due west of turbine B.

a Calculate the distance AC. **(3 marks)**

b Calculate the bearing of C from A.

Give your answer correct to the nearest degree. **(4 marks)**

June 2014, Q15, 1MA0/2H

14 Problem-solving OABC is a trapezium.

$\overrightarrow{OA} = \mathbf{a}$

$\overrightarrow{OB} = \mathbf{b}$

$\overrightarrow{BC} = 2\mathbf{a}$

Write in terms of **a** and **b**:

a $\overrightarrow{OC}$

b $\overrightarrow{BA}$

c $\overrightarrow{AC}$

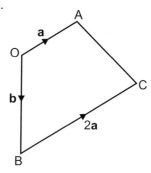

15

The diagram shows a solid made from a hemisphere and a cone.

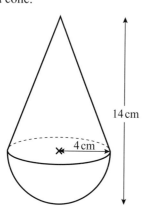

14 cm

4 cm

Diagram **NOT** accurately drawn

The radius of the hemisphere is 4 cm.

The radius of the base of the cone is 4 cm.

Calculate the volume of the solid.

Give your answer correct to 3 significant figures.   **(3 marks)**

November 2014, Q8, 1MA0/1H

Exam hint

Work out the height of the cone.

16 Reasoning Lines EG and FH are parallel.

Lines AB and CD are parallel.

Angle EAB = 100°

Angle BCD = 50°

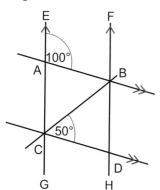

a   Show that triangle ABC is isosceles.

b   Show that triangle BCD is congruent to triangle ABC.

c   Hence show that ABCD is a rhombus.

Q16c hint Hence means 'use your answer to the previous part'.

6 STATISTICS

What is covered in this section	Foundation Student Book reference
6.1 Scatter graphs • Draw scatter graphs. • Determine whether or not there is a relationship between sets of data. • Draw a line of best fit and use it to predict values. • Identify outliers.	Unit 3
6.2 Mean, median, mode and range • Estimate the mean and range for a set of grouped data. • Find the modal class and the class containing the median.	Unit 7
6.3 Sampling • Identify sources of bias in sampling. • Describe how to select a random sample. • Use stratified sampling.	Unit 7
6.4 Mixed exercise • Consolidate your learning with more practice.	

6.1 Scatter graphs

Objectives

- Draw scatter graphs.
- Determine whether or not there is a relationship between sets of data.
- Draw a line of best fit and use it to predict values.
- Identify outliers.

Key point 1

A **scatter graph** shows the relationship between two sets of data. Plot the points with crosses. Do not join them up.

Key point 2

The relationship between the sets of data is called **correlation**. The sets of data are called **variables.**

Positive correlation 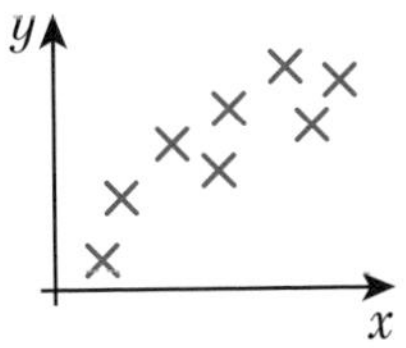
As x increases
y increases

Negative correlation 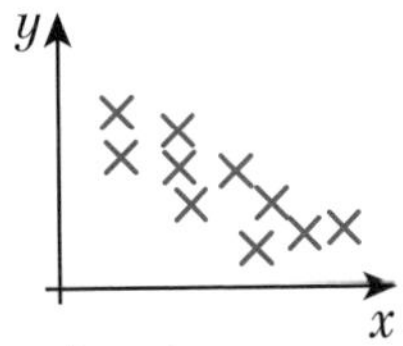
As x increases
y decreases

No correlation 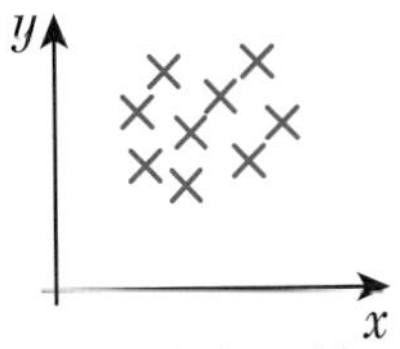
No relationship
between x and y

75

Key point 3

An **outlier** is a value that does not fit the pattern of the data.

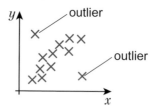

An outlier can be ignored if it is due to a measuring or recording error.

Key point 4

A **line of best fit** is a straight line drawn through the middle of the points on a scatter graph. It should pass through as many points as possible and represent the trend of the points.

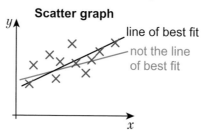

Key point 5

Using a line of best fit to predict data values within the range of the data given is called **interpolation** and is usually reasonably accurate.

Using a line of best fit to predict data values outside the range of the data given is called **extrapolation** and may not be accurate.

1   The scatter graph shows the petrol consumption and engine size of different cars.

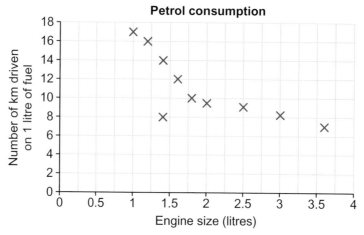

a   What type of correlation does it show?

b   Copy and complete this sentence to describe the relationship between engine size and number of kilometres driven on one litre of fuel.
The larger the engine size, the _____ the number of kilometres driven on one litre of fuel.

c   Fred thinks the engine size was recorded wrongly for one of the cars.
How many km did this car drive on one litre of fuel?

2 The table shows the heights and weights of some athletes.

Q2 A common error is to think the line of best fit must go through (0, 0).

Height (cm)	171	161	165	158	184	176	191
Weight (kg)	64	58	64	51	74	75	82

  a Draw a scatter graph to show this data.
  b Describe the correlation.
  c Draw a line of best fit.
  d Use your line of best fit to predict the weight of an athlete who is 180 cm tall.

3 **Reasoning** An elastic rope is suspended from a beam and weights are hung on its end. The table shows the length of the rope for different weights.

Weight (N)	2	3	3.5	4	5	5.8	6.5	7.1	7.5	8.3
Length (cm)	30	31	38	32	38	42	40	17	44	52

  a Draw a scatter graph to show this data.
  b Circle the outlier on your graph. Explain why this value can be ignored.
  c Draw a line of best fit.
  d Use your line of best fit to estimate the weight that would stretch the rope to 40 cm.
  e Estimate the length of the rope with no weights on it.

4 **Problem-solving** The table shows the air pressure (in pascals, Pa) recorded at different heights above sea level.

Height above sea level (m)	0	500	1200	1500	2000	3000	4000	5000
Air pressure (Pa)	101 000	96 000	88 000	85 000	81 000	72 000	63 000	56 000

  a Draw a scatter graph to show this data.
  b Draw a line of best fit.
  c Estimate the air pressure at 1800 m.
  d The air pressure outside an aeroplane is 70 000 Pa. At what height is it flying?

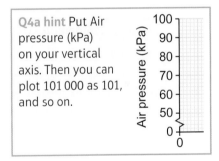

Q4a hint Put Air pressure (kPa) on your vertical axis. Then you can plot 101 000 as 101, and so on.

5 **Problem-solving** The ages and IQ scores of 10 people are given in the table.

Q5 A common error is to try to work out if there is a correlation without drawing a scatter graph.

Age (years)	30	24	50	54	40	63	24	39	44	118
IQ score	90	104	87	92	100	112	120	113	95	56

Does this data show a correlation between age and IQ score?

6 **Reasoning** The table shows the lengths and weights of 10 beetles, all from one species.

Length (mm)	4.6	5.8	4.3	4.9	5.6	4.8	5.1	5.1	4.7	5.0
Weight (mg)	24	35	22	30	33	28	31	33	27	29

  a Does the data show a relationship between length and weight for these beetles?
  b Another two beetles are captured and measured.
     Beetle A   length 4.7 mm   weight 27 mg
     Beetle B   length 5.2 mm   weight 23 mg
     Is each of these beetles likely to be from the same species as the 10 in the table? Explain.

7 **Reasoning** The scatter graph shows the heights of a group of people and the number of films they each watched last month.

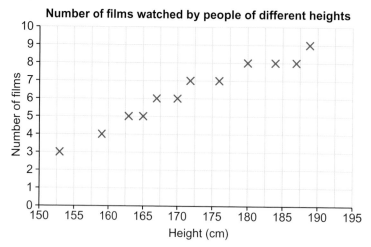

Number of films watched by people of different heights

a Does the graph show a correlation between height and number of films watched?

b Does the graph show causation between height and number of films watched?

Q7b hint Does being taller make you watch more films?

Q7b communication hint **Causation** means that a change in one variable *causes* a change in the other variable.

8 **Problem-solving** Arthur recorded his heating costs and the average daily temperature.

Month	Jan	Feb	Mar	Apr	May
Average temperature (°C)	4	6	9	12	14
Cost (£)	85	78	62	44	35

a Draw a scatter graph for his data.

b Estimate the heating cost for a month when the average temperature is
   i 20 °C   ii 10 °C

c Which of your estimates in part **b** is likely to be more reliable? Explain.

Q8b hint Draw a line of best fit.

9 **Reasoning** a Which of these graphs shows the stronger correlation? Explain how you know.

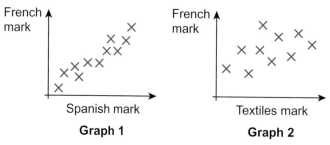

Graph 1

Graph 2

b Do you think either graph shows causation? Explain.

Q9b hint Do you think being good at one of the subjects could make you better at the other?

10 **Reasoning** Match each example to the type of correlation.
   a Temperature and sales of gloves
   b Arm length and number of siblings
   c Weight of parcel and cost of postage

   A No correlation
   B Positive correlation
   C Negative correlation

6.2 Mean, median, mode and range

Objectives

- Estimate the mean and range for a set of grouped data.
- Find the modal class and the class containing the median.

Key point 6

For grouped data, you can use the midpoint of each class to calculate an **estimate** for the mean.

Key point 7

To estimate the range of grouped data, work out
   maximum possible value – minimum possible value

Key point 8

The **modal class** is the class with the highest frequency.

Key point 9

For a set of n values, the $\frac{n+1}{2}$ th value is the median.

---

1   A forester measured the heights of 50 trees. The table shows the results.

Height, h (m)	Frequency, f	Midpoint of class, m	$m \times f$
$0 < h \leqslant 2.0$	8		
$2.0 < h \leqslant 4.0$	12		
$4.0 < h \leqslant 6.0$	15		
$6.0 < h \leqslant 8.0$	7		
$8.0 < h \leqslant 10.0$	8		
Total		**Total**	

a   Copy and complete the frequency, midpoint and $m \times f$ columns.

b   Calculate an estimate for the mean height.

> **Q1b hint** $\dfrac{\text{Total of } m \times f}{\text{Total } f}$

2   A botanist counted the numbers of seeds in 30 pods from one plant species.
The table shows the results.

Number of seeds	Frequency
1–5	5
6–10	12
11–15	10
16–20	3

> **Q2 hint** Copy the table and add the extra columns and totals, as in **Q1**.

Calculate an estimate for the mean number of seeds per pod. Give your answer to 1 d.p.

3   There are 100 chickens on a farm.
The table shows how many eggs they lay each week.
Calculate an estimate for

a   the mean   b   the range.

Number of eggs	Frequency
0–1	12
2–3	22
4–5	45
6–7	21

> **Q3** A common error is to divide by the number of classes (4 here) instead of the total frequency.

4   The table shows the lengths of calls to a call centre, timed to the nearest minute.

Length of call, t (min)	Frequency
$0 < t \leqslant 2$	36
$2 < t \leqslant 6$	41
$6 < t \leqslant 10$	58
$10 < t \leqslant 20$	15

a   How many calls were there?

b   Which is the modal class?

Q4a hint What is the total frequency?

c   Calculate an estimate for the mean time, to the nearest minute.

d   How many calls were longer than the mean time?

5   **Problem-solving** A scientist recorded the heights of some plants, to the nearest cm.

Length, l (cm)	Frequency
$0 < l \leqslant 5$	7
$5 < l \leqslant 10$	12
$10 < l \leqslant 15$	15
$15 < l \leqslant 20$	5
$20 < l \leqslant 25$	1

She calculated

• estimate of mean = 15.7

• estimate of range = 25

a   Without doing any calculations, do you think her
    estimate for the mean is correct?
    Explain.

Q5a hint Does her mean represent most of her values?

b   If her mean is incorrect, calculate a better estimate.

c   Explain why her results are both estimates.

6   **Reasoning** The table shows the speeds of cars recorded by a speed camera
    in a 40 mph zone.

Speed, s (mph)	Frequency
$20 < s \leqslant 25$	24
$25 < s \leqslant 30$	36
$30 < s \leqslant 35$	20
$35 < s \leqslant 40$	71
$40 < s \leqslant 45$	37
$45 < s \leqslant 50$	12

a   How many cars were breaking the 40 mph speed limit?

b   Which is the modal class?

c   Which class contains the median?

d   Calculate an estimate for the mean speed.

e   Local residents say the average speed is
    faster than the 40 mph speed limit.
    Are they correct?
    Explain.

Q6d hint Midpoint of
$20 < s \leqslant 25$ is $\dfrac{20 + 25}{2} = \dfrac{45}{2} = \square$

7 **Reasoning** The table shows test results from one maths group.

Mark	Frequency
0–20	1
21–40	3
41–60	8
61–80	7
81–100	6

Shona says the average mark is between 41 and 60.
Pritti says the average mark is between 61 and 80.

a   Which average is Shona using?

b   Which average could Pritti be using?

c   Estimate the range.
    Why is this value probably an over-estimate?

Q7 hint Calculate the midpoints carefully.
$$\frac{0+20}{2} = \square \qquad \frac{21+40}{2} = \square$$

Q7c hint Think about the likely highest and lowest marks.

8 **Problem-solving** On one day each month, a manager at a health centre records the times patients wait to see a doctor. Last month, the mean waiting time was 7.2 minutes. The table shows the data for the next month. Had the mean waiting time gone down?

Waiting time, t (min)	Frequency
$0 < t \leqslant 5$	36
$5 < t \leqslant 10$	35
$10 < t \leqslant 15$	14

6.3 Sampling

Objectives

- Identify sources of bias in sampling.
- Describe how to select a random sample.
- Use stratified sampling.

Key point 10

In a survey, a **sample** is taken to represent the **population**. A sample that is too small can **bias** the results.

Key point 11

In a **random sample**, every member of the population has an equal chance of being included.

Key point 12

The population may be divided into groups, e.g. men and women.
A **stratified sample** contains members of each group in proportion to the size of the group.

Key point 13

In order to reduce **bias**, a sample must – as far as possible – represent the whole population being considered.

1   A newspaper editor wants to find out people's opinions about the new design of the newspaper. The newspaper has 80 000 readers.

a   Explain why the editor should take a sample.

b   What size sample would be the most suitable?
    100   400   6000

c   The editor decides to take an 8% sample. How many people will be in the sample?

2   Darren wants to find out about people's shopping habits. He interviews people in the city centre shopping area on a Saturday. Is his sample likely to be representative of the population? Explain.

Q2 hint Think about the time and place of the survey.

3   A sports centre wants to run new classes aimed at 16–21-year-olds. They plan to pick 1000 names at random from the local phone book and ask these people which classes they should run. Explain why this sample could be biased.

4   **Reasoning** A form tutor wants to pick 5 students at random as school council representatives. Which of these methods will give the best sample? Explain.
    A   Pick the top 5 names from the register.
    B   Give each student a raffle ticket, then pick 5 tickets from a hat.
    C   Pick the last 5 students to leave one afternoon.

Q4 hint Explain why the method you pick is best, and what is wrong with the other two.

5   Describe one way of picking a random sample of 20 people from 200 people at a cinema.

Q5 hint You could use seat numbers or ticket numbers.

6   **Reasoning** Explain whether each of these samples is biased.
    a   A dentist's surgery wants to find out what its patients think about their service.
        They ask everyone who comes into the surgery one Tuesday.
    b   A market research company wants to find out how much people spend each week on food.
        It carries out a telephone survey of people in four towns.
    c   You want to find out the most popular type of music and ask 50 people at a rock concert.
    d   A doctor's surgery wants to find out what its patients think about their service.
        They select names from their patient list at random and send questionnaires to them.

Q6 A common error is to not explain clearly why a sample could be biased.

7   Out of 30 people at a martial arts class
    • $\frac{2}{3}$ are male.   • $\frac{1}{3}$ are female.
    Tilly wants to take a stratified sample of 12 people.
    a   What fraction of her sample should be male?
    b   What fraction of her sample should be female?
    c   Work out the number of males and the number of females in her sample.

Q7 hint The fractions in the sample should be the same as in the population.

8   There are 400 students at a college. 240 of them are in Year 12.
    a   What fraction of the students are in Year 12?
    b   The Principal wants to take a stratified sample of the students.
        There will be 80 students in the sample.
        How many Year 12 students should there be in the sample?

9   **Problem-solving** A company has 24 part-time and 36 full-time members of staff.
    They want to take a stratified sample of 15 employees to find out their views on flexi-time.
    How many of the employees in the sample should be
    a   part-time   b   full-time?

Q9a hint What fraction of the employees are part-time?

Q9 A common error is to not check the final numbers add up to the sample size you want.

10 **Problem-solving** The owner of a stately home wants to find out whether visitors think their tickets are good value for money.
    In one month
    • 3000 visitors buy house and garden tickets
    • 5000 visitors buy garden-only tickets.
    a   He wants to take a 5% sample. How many people does he need?
    b   He wants to take a stratified sample by ticket type. How many does he need for each ticket type?
        i   house and garden   ii   garden only

6.4 Mixed exercise

Objective

• Consolidate your learning with more practice.

1 **Reasoning** Deepak is doing a survey to find out how often people travel by train.
He is going to ask 10 men leaving a bus station.
Give two reasons why this may not produce a good sample for his survey.

2 **Reasoning** A political party is doing a survey to find out how people might vote in an election.
Which of these methods would give them a more representative sample?
A Online survey via Facebook
B Pick names at random from the electoral roll and send these people a paper survey.

3 **Exam question**

Alice is a lorry driver.

She recorded the distance she drove on each of 40 trips.

The table gives information about these distances.

Distance, d (miles)	Frequency
$400 < d \leqslant 450$	9
$450 < d \leqslant 500$	15
$500 < d \leqslant 550$	12
$550 < d \leqslant 600$	4

Work out an estimate for the mean distance. **(4 marks)**

November 2013, Q8, 5MB1H/01

Q3 A common error in this question was to use the interval (50) instead of the midpoint of each interval.

4 **Exam question**

Carlos has a cafe in Clacton.

Each day, he records the maximum temperature in degrees Celsius (°C) in Clacton and the number of hot chocolate drinks sold.

The scatter graph shows this information.

On another day the maximum temperature was 6°C and 35 hot chocolate drinks were sold.

a Show this information on the scatter graph.

b Describe the relationship between the maximum temperature and the number of hot chocolate drinks sold.

c Draw a line of best fit on the scatter diagram.

One day the maximum temperature was 8 °C.

d Use your line of best fit to estimate how many hot chocolate drinks were sold. **(4 marks)**

June 2014, Q2, 5MB1H/01

Hot chocolate drinks sold / Maximum temperature (°C)

Exam hint

Draw lines from the axes to your line of best fit, to show your method in part **d**.

5 **Problem-solving** The table shows the amount of rainfall on 30 days in an English town.

a Which is the modal class?

b Which class contains the median?

c Draw a histogram to show this data.

d In what season do you think this data was collected? Explain.

Rainfall, r (mm)	Frequency
$0 \leq r < 2$	18
$2 \leq r < 4$	6
$4 \leq r < 6$	2
$6 \leq r < 8$	3
$8 \leq r < 10$	1

6 **Exam question**

The scatter graph shows information about 10 newborn babies. It shows each baby's body length and head circumference.

Another baby has a body length of 47 cm and head circumference 34 cm.

a Show this information on the scatter graph.

b What type of correlation does the scatter graph show?

A baby has head circumference 35 cm.

c Estimate the body length of this baby. **(4 marks)**

November 2014, Q9, 5MB1H/01

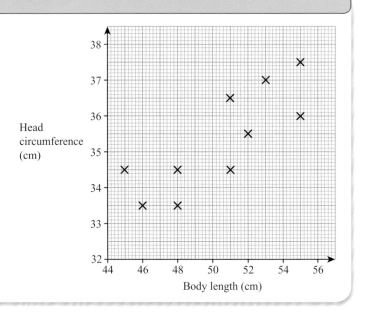

Exam hint

Students who did not draw a line of best fit were more likely to get the answer to part **c** wrong.

7 **Problem-solving** The table shows the annual salaries for 2482 employees in a large company.

Salary, s (£)	Frequency
$15\,000 \leq s < 20\,000$	112
$20\,000 \leq s < 25\,000$	328
$25\,000 \leq s < 30\,000$	639
$30\,000 \leq s < 40\,000$	563
$40\,000 \leq s < 50\,000$	482
$50\,000 \leq s < 60\,000$	261
$60\,000 \leq s < 90\,000$	97

a Which is the modal class?

b Which class contains the median?

c Calculate an estimate for the mean salary, to the nearest £.

d The company is trying to attract new trainees. Its advertisements say:

On average, our employees earn £_____

Which of the averages from parts **a**, **b** and **c** should the company use in this advertisement?

> **Q7b hint** For large data sets, use $\frac{n}{2}$ to find the position of the median.

8   **Problem-solving** The histogram shows the waiting times for rides in a theme park.

Waiting times for rides

a   Which is the modal class?
b   Copy and complete the frequency table.
c   Calculate an estimate of the mean waiting time.
d   Find the class that contains the median.
e   Estimate the range.

Time, t (min)	Frequency
$0 \leqslant t < 5$	

Q8 c, d, e hint Use the frequency table.

9   At a golf club, the ratio of men to women is $4 : 1$.
The club secretary wants to survey a sample 30 members
to find out their views on the club's opening times.
How many women should there be in the sample?

Q9 hint Divide the sample in the same ratio as the club.

10 **Problem-solving** The histogram shows the heights of
sunflowers grown from seeds in one packet.
On the packet it says 'Average height over 1.7 metres'.
Is this claim correct?
Explain.

Q10 hint Make a frequency table for the data.

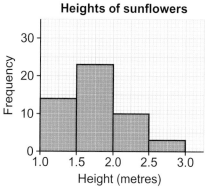

Heights of sunflowers

11 **Problem-solving** The frequency table shows the ages of 100 people at a health club.

Age, a (years)	Frequency
$20 < a \leqslant 30$	25
$30 < a \leqslant 40$	23
$40 < a \leqslant 50$	18
$50 < a \leqslant 60$	24
$60 < a \leqslant 70$	10

a   Calculate an estimate for the mean age, to the nearest year.
b   Which class contains the median age?

The manager wants to pick a sample of 20 people, stratified by age.
c   How many people in the age range $60 < a \leqslant 70$ should there be in the sample?
d   How many people in the age range $20 < a \leqslant 30$ should there be in the sample?

85

ANSWERS

1 NUMBER

1.1 Calculations

1 a 500   b 20   c 9   d 8

2 a $\frac{2+10}{6}=2$ b $\sqrt{80+20}=10$

3 a $675 \leqslant n < 685$ b $1.265 \leqslant n < 1.275$
   c $17.15 \leqslant n < 17.25$ d $195 \leqslant n < 205$

4 $1.995 \leqslant l < 2.005$

5 Minimum: 228 ml; maximum: 252 ml

6 No, because 115 g is less than the lowest possible value, which is 116.4 g.

7 $900 \div 20 = 45$

8 a $\frac{10+9}{6-2} \approx \frac{20}{4} = 5$; could be correct
   b $\sqrt{\frac{4+60}{16}} = \sqrt{\frac{64}{16}} = 2$; wrong
   c $\frac{5000-6400}{5+10} = \frac{-1400}{15} \approx -100$; wrong
   d $\frac{1000 \times 4}{2 \times 5^2} = \frac{4000}{50} = 80$; could be correct

9 b 2.02 (3 s.f.)   c −70.3 (3 s.f.)

10 a 1.4 (2 s.f.)   b 0.960 (3 s.f.)   c 4.95 (3 s.f.)   d 1.457 (4 s.f.)

1.2 Factors and multiples

1 $2^3 \times 3 \times 5^2$

2 a 4   b 30

3 a

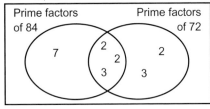

   b $84 = 2^2 \times 3 \times 7$, $105 = 3 \times 5 \times 7$

4 $72 = 2^3 \times 3^2$

5 HCF = 12, LCM = 504

Prime factors of 84 | Prime factors of 72

7 | 2 2 3 | 2 3

6 a HCF = 21, LCM = 420   b HCF = 14, LCM = 140
   c HCF = 12, LCM = 660

7 24 and 40

8 10 and 35

9 a i $2 \times 7 = 14$ ii 4   iii 3
   b i A and C ii B and C iii A and D
   c B and C

10 10:36 am

1.3 Fractions

1 a $\frac{43}{30} = 1\frac{13}{30}$ b $\frac{10}{36} = \frac{5}{18}$ c $\frac{15}{28}$ d $\frac{16}{9} = 1\frac{7}{9}$

2 a $\frac{5}{14}$ b $\frac{3}{14}$ c $\frac{1}{3}$ d $\frac{1}{6}$

3 a $3\frac{3}{5}$ b $5\frac{26}{35}$ c $6\frac{25}{36}$ d $5\frac{19}{30}$

4 a $4\frac{3}{10}$ b $8\frac{3}{8}$ c $8\frac{11}{21}$ d $5\frac{17}{36}$

5 a $4\frac{11}{45}$ b $2\frac{1}{8}$ c $3\frac{11}{21}$ d $1\frac{18}{55}$

6 a $2\frac{13}{20}$ b $\frac{25}{28}$ c $2\frac{8}{9}$ d $1\frac{3}{4}$

7 a $2\frac{2}{15}$ b $\frac{9}{10}$ c $2\frac{1}{2}$ d $1\frac{4}{5}$

8 a $11\frac{1}{4}$ b $2\frac{5}{8}$ c $4\frac{2}{5}$ d 3

9 a $2\frac{11}{12}$ b $1\frac{7}{8}$ c $4\frac{2}{7}$ d 8

10 a $1\frac{5}{28}$ b $1\frac{1}{35}$ c $\frac{5}{8}$ d $1\frac{1}{2}$

11 Yes, because the total is $8\frac{53}{120}$ kg.

12 a 5 lengths   b $1\frac{1}{10}$ m

1.4 Indices, powers and roots

1 a 2187   b 1024   c 625   d x^2
   e $\frac{1}{16}$ f $\frac{8}{27}$ g $\frac{1}{y^5}$ h $\frac{x^2}{25}$

2 a 4   b 3   c $\frac{1}{2}$ d $\frac{2}{3}$

3 a 5^6 b 2^{12} c 6^9 d 10^{10}
   e x^8 f t^{12} g m^8 h n^6

4 a 2^{14} b 4^{10} c 5^7 d 10^6
   e 3   f 2^{12} g x^{16} h n^4

5 a $\frac{1}{3}$ b $\frac{1}{2}$ c $\frac{1}{10}$ d $\frac{1}{y}$
   e 4   f $\frac{5}{2}$ g z h $\frac{b}{a}$

6 a 1   b 1   c $\frac{1}{2}$ d 1
   e $\frac{1}{5}$ f 9   g 36   h $\frac{1}{10}$

7 a 1   b $\frac{1}{t}$ c $\frac{1}{n}$ d m

8 $6^{-1}, 2^{-1}, 8^0, \left(\frac{2}{5}\right)^{-1}, \left(\frac{1}{4}\right)^{-1}$

9 Parts **b**, **c**, **e**, **f** and **h**

10 a 2   b 5   c 7   d n
   e 12   f 15   g 24

11 a $7\sqrt{3}$ b $\sqrt{5}$ c $6\sqrt{17}$ d $2\sqrt{11}$

12 a $\frac{1}{16}$ b $\frac{1}{243}$ c $\frac{1}{125}$ d $\frac{1}{49}$
   e $\frac{1}{100}$ f 9   g 16   h 27
   i $\frac{25}{9} = 2\frac{7}{9}$ j $\frac{49}{4} = 12\frac{1}{4}$ k $\frac{8}{125}$

13 a n^2 b $\left(\frac{d}{c}\right)^3$ c $\frac{1}{x^2}$ d $\frac{1}{r^2}$
   e t^5 f $\frac{1}{x^4}$ g y h $\frac{1}{z^2}$

14 a $m = 5$ b $c = 4$ c $d = -2$ d $x = 2$
   e $y = -1$ f $n = 2$ g $z = -2$

1.5 Standard form

1 a 10^8 b 10^5 c 10^2 d 10^{-6}

2 a 0.001   b 0.01   c 100 000   d 0.1

3 a 10^{-4} b 10^3 c 10^{-5} d 10^6
   e 10^{-6} f 10^4

4 a 700 000   b 160 000   c 23 700   d 362.05
   e 5 290 260   f 419 030   g 976   h 126 000

5 a 2×10^3 b 8×10^5 c 5.6×10^3
   d 9.265×10^4 e 1.254×10^5 f 1.253×10^2
   g 2.1591×10^7 h 5.4×10^6

6 a 5.2×10^{-3} b 4.31×10^{-6} c 6.59×10^{-3}
   d 4.71×10^{-4} e 5.0802×10^{-3} f 9.5×10^{-5}
   g 1.4×10^{-7} h 2.37×10^{-2}

7 a 0.000 027   b 0.003 46
   c 0.092 01   d 0.000 84
   e 0.000 000 000 506   f 0.000 001 72
   g 0.000 000 0654   h 0.000 000 5192

8 a Yes   b 3.26×10^{-1} c 1.763×10^{-1} d 1.05×10^4

9 a 6×10^7 b 6×10^4 c 3.2×10^2 d 4×10^8
   e 3.5×10^6 f 1.5×10^2 g 3.6×10^{-5} h 2.1×10^{-2}

10 a 3×10^3 b 3.5×10^{-5} c 2×10^6 d 2×10^{-1}
   e 2.66×10^4 f 3.2×10^2 g 1.94×10^2 h 4.87×10^2

11 a 3.4×10^4 b 7.5×10^{-2} c 6.5×10^4
   d 2.95×10^{-2} e 8.256013×10^3 f 5.131×10^6
   g 3.559×10^{-2} h 2.75031×10^7

12 a 1.7×10^9 b 4.4×10^{-2} c 4.0944×10^2
     d 7.065×10^8 e 4.189×10^{-2} f 8.4599908×10^8
     g 5.258×10^{-3} h 9.7519×10^{15}

1.6 Mixed exercise

1   11 am

2 $\frac{30 \times 10}{0.5} = 600$ (not 601.11984…)

3 $\frac{20 \times 300}{0.5} = 12\,000$ bottles

4 a $42 = 2 \times 3 \times 7$, $504 = 2^3 \times 3^2 \times 7$
    b Yes, because $2^3 \times 3^2 \times 7$ is a multiple of $2 \times 3 \times 7$

5 i 3 packets and 5 boxes (or 6 and 10, or 9 and 15, etc.)
    ii 60 bread rolls (or a multiple of 60)

6   Bolt C, because $4\frac{1}{4} > 2\frac{5}{8} + 1\frac{3}{10} = 3\frac{37}{40}$

7 a 55   b $14 : 6 : 21$

8 $9\frac{31}{32}$ square yards

9 a 3×7 b 3×5 c 2×7

10 $3\frac{25}{26}$ feet

11   No, because 500 sheets are 4.5 cm deep (or because $4 \div 500 = 0.008$ and $0.008 < 9 \times 10^{-3}$)

12   About 20 times bigger

13   About 400 songs

14 a 8×10^6 b 6.4×10^{16} c 2.7×10^{-8} d 1.25×10^{-1}

15   Parts **b**, **e** and **f**

16 $2 \times 10^{10} \, \text{m}^2$

17   about 36 prizes

18 0.38×10^{-1}, 3800×10^{-4}, 0.038×10^2, 380

19 a $\frac{1}{5}$ b $\frac{1}{9}$ c 2.7×10^8

20 a $2\frac{2}{49}$ b $7\frac{21}{25}$ c $2\frac{10}{27}$ d $1\frac{61}{64}$

21 3.92×10^{18} km, 4.08×10^{18} km

22 a $7\sqrt{2}$ b $\sqrt{39}$ c $\sqrt{10}$ d $2\sqrt{5}$

2 ALGEBRA

2.1 Writing equations and formulae

1 a $P = 2x + 10$ b $P = 7a - 10$ c $P = 3y + 19$

2 a $7n + 4 = 26$; $n = 3$ b $\frac{n}{3} - 2 = 2$; $n = 12$
    c $5(n - 4) = 30$; $n = 10$ d $\frac{n + 8}{2} = 7$; $n = 6$

3 a $P = 7a + 10b$ b £41

4 $W = nx + m(x + 4)$

5   35

6   12

7   15, 7, 28

8   20 cm

9 a $P = 4x + 12$ b $A = x^2 + 6x$

10 $A = lw + \frac{1}{2}lh$

11 a Ryan   b $A = 2lx + lw + wh$

2.2 Rearranging formulae

1 a $x = \frac{y}{m}$ b $r = tv$ c $w = \frac{l}{T}$ d $t = \frac{x - y}{2}$
    e $s = r - 5q$ f $h = \frac{m + n}{6}$ g $y = 3x - 2z$ h $c = \frac{3a + 2b}{5}$

2 a $-2x - 3$ b $-\frac{1}{3}$ c $-x + y$ d $-2x - 1$

3 a $x = \frac{z - y}{3}$ b $x = \frac{3c - d}{2}$ c $x = \frac{6e + 7}{4}$ d $x = \frac{fg - h}{5}$
    e $x = 3m - kl$ f $x = \frac{np - m}{2}$ g $x = \frac{t - rs}{p}$ h $x = \frac{aw - uv}{z}$

4 a $y = \frac{A}{xz}$ b $t = \frac{v}{5s}$ c $q = \frac{m}{pr}$ d $u = \frac{3s}{4t}$
    e $x = \frac{4z}{y}$ f $a = \frac{3w}{b}$ g $f = \frac{5d}{e}$ h $r = \frac{2p}{s}$
    i $k = \frac{5g}{3h}$ j $a = \frac{7t}{4x}$ k $n = \frac{mx}{p}$ l $b = \frac{kt}{ax}$

5 a $y = \frac{x + 2z}{2}$ or $y = \frac{x}{2} + z$ b $y = \frac{m - 5t}{5}$ or $y = \frac{m}{5} - t$
    c $y = \frac{4k - d}{4}$ or $y = k - \frac{d}{4}$ d $y = \frac{r - ac}{a}$ or $y = \frac{r}{a} - c$
    e $y = \frac{3n - 5z}{5}$ or $y = \frac{3n}{5} - z$ f $y = \frac{2r - 5q}{2}$ or $y = r - \frac{5q}{2}$
    g $y = \frac{6x}{k} + n$ h $y = \frac{4l + n}{n}$ or $y = \frac{4l}{n} + 1$

6   C, A, B

7 a $x = at - 2$ b $x = bv + 3$
    c $x = 3(n - m)$ d $x = t(y + d)$
    e $x = y(z - 2) - 4$ f $x = b(a - 3) + 5$
    g $x = i(h + 7) + k$ h $x = 6 - r(p - 4)$

8 a $y = \sqrt{\frac{x}{b}}$ b $z = \sqrt{\frac{m}{k}}$ c $a = \frac{r}{t^2}$
    d $b = \sqrt{c^2 - a^2}$ e $n = m^2$ f $r = \frac{t^2}{2}$
    g $b = \sqrt{\frac{R}{ac}}$ h $t = \frac{s^2}{\pi}$

9 a $\frac{z}{5 - y}$ b $x = \frac{t}{4 - y}$ c $x = \frac{m}{r + 3}$ d $x = \frac{s}{5 - t}$
    e $x = \frac{b}{a + 6}$ f $x = \frac{d - 4}{c + 2}$ g $x = \frac{t + 3}{m - n}$ h $x = \frac{-p - s}{u - r}$

10 a $t = \frac{10 - a}{7}$ b $b = \frac{c + 24}{5}$ c $a = \frac{d + 3m}{m - 1}$ d $r = \frac{x - 4n}{n - 1}$
    e $d = \frac{6 + 5e}{3}$ or $d = 2 + \frac{5e}{3}$
    f $r = \frac{4p + 7}{3}$ g $x = \frac{2m + r}{r - m}$ h $b = \frac{5d - 3a}{a - d}$

11 a $b = \sqrt{a^2 - c^2}$ (C)   b $b = (a - c)^2$ (B)
    c $b = a^2 - c$ (A)   d $b = \sqrt{a} - c$ (D)

2.3 Expanding and factorising

1 a $x^2 + 5x + 2x + 10 = x^2 + 7x + 10$
    b $x^2 + 4x + 3x + 12 = x^2 + 7x + 12$
    c $x^2 + 2x - 3x - 6 = x^2 - x - 6$

×	x	+2
x	x^2	+2x
+5	+5x	+10

2   Expressions **a**, **b**, **e**, **f**

3 a $x^2 + 6x + 8$ b $x^2 + 11x + 30$
    c $a^2 + 10a + 16$ d $m^2 + 15m + 54$
    e $t^2 + 12t + 35$ f $r^2 + 11r + 28$

4 a $x^2 + 3x - 18$ b $x^2 + 7x - 8$
    c $y^2 - 2y - 24$ d $n^2 - 6n - 7$
    e $x^2 + 5x - 36$ f $x^2 + 8x - 33$

5 $(x - 2)(x - 5) = x^2 - 7x + 10$
    $(x - 2)(x - 1) = x^2 - 3x + 2$
    $(x - 2)(x - 4) = x^2 - 6x + 8$
    $(x - 5)(x - 4) = x^2 - 9x + 20$
    $(x - 5)(x - 1) = x^2 - 6x + 5$
    $(x - 4)(x - 1) = x^2 - 5x + 4$

6 a $x^2 + 6x + 9$ b $m^2 - 4m + 4$
    c $d^2 + 18d + 81$ d $r^2 + 20r + 100$
    e $x^2 - 10x + 25$ f $y^2 + 22y + 121$
    g $z^2 - 12z + 36$ h $x^2 - 14x + 49$

7 a $x^2 - 1$ b $x^2 - 25$ c $x^2 - 49$ d $x^2 - 144$

8 $b^2 - 9$

9 a $(x - 3)(x + 3)$ b $(x - 6)(x + 6)$
    c $(t - 4)(t + 4)$ d $(h - 10)(h + 10)$

10 a $x^2 + 8x + 16$ (D)   b $x^2 + 6x + 8$ (C)
    c $x^2 + 10x + 16$ (B)   d $x^2 + 4x + 4$ (A)

11 a $(x + 6)(x + 7)$ b $(x + 8)(x + 3)$ c $(x + 5)(x + 1)$
    d $(x + 1)^2$ e $(y + 2)(y + 7)$ f $(z + 1)(z + 9)$

12 a $(x - 5)(x + 3)$ b $(x - 4)(x + 5)$ c $(n - 9)(n + 2)$
    d $(t + 7)(t - 8)$ e $(y - 3)(y + 7)$ f $(x - 6)(x + 2)$

13 a $(x - 7)(x - 2)$ b $(x - 3)(x - 2)$ c $(x - 4)(x - 8)$
    d $(f - 5)(f - 9)$ e $(v - 8)(v - 10)$ f $(x - 1)(x - 6)$

14 a $(x - 8)(x + 8)$ (C)   b $(x + 8)^2$ (D)
    c $(x - 8)^2$ (A)   d $(x + 4)(x + 16)$ (B)

15 $20(x + 5)(x - 3) = 20(x^2 + 2x - 15)$ or $20x^2 + 40x - 300$

2.4 Inequalities

1 a $x < 1$　　b $x \geqslant 2$　　c $-4 < x \leqslant 2$
　 d $2 \leqslant x < 10$　 e $-20 \leqslant x \leqslant 11$　f $45 < x < 55$
2 a 1, 2, 3, 4, 5, 6　 b 1, 2, 3　　　c -1, 0, 1, 2, 3
　 d -2, -1, 0, 1　 e 0, 1, 2, 3　　f -12, -11, -10, -9
3 a $x < 6$　　b $x \geqslant 10$　c $x \geqslant -3$　d $x < 7$
　 e $x > 1$　　f $x \geqslant -4$　g $x \geqslant 7.7$　h $x < 1\frac{1}{2}$
4 a $x \geqslant 4$　　b $x < 7$　　c $x \leqslant 2$　　d $x > -3$
　 e $x \leqslant \frac{9}{6}$ or $\frac{3}{2}$　f $x \geqslant \frac{1}{2}$　g $x > 2$　　h $x < 0.5$
5 a $x < 10$　　b $x \leqslant 18$　c $x \geqslant 20$　d $x > 10$
　 e $x \leqslant \frac{3}{4}$　　f $x > 2$　　g $x > 35$　h $x \geqslant \frac{1}{12}$

6 a $x > 5$

　 b $x \leqslant 3$

　 c $x < 6$

　 d $x \geqslant 5$

　 e $x \leqslant -2$

　 f $x < 9$

　 g $x \geqslant 28$

　 h $x \geqslant -6$

7 a $x > 3$　　b $x \leqslant 6$　c $x < 14$　d $x \geqslant 13$
　 e $x > -4$　f $x \leqslant 10$　g $x > 4.5$　h $x < 13$
8 a $x < -3$ (D)　b $x > 3$ (B)　c $x < -5$ (A)　d $x < 3$ (C)
9 a $x \leqslant 1$　　b $x > -15$　c $x > -5$　d $x \leqslant -1$
　 e $x < 3$　　f $x \geqslant 3$　　g $x < 3.5$　h $x > 9.5$
10 $x < 3.5$, so $x = 1$, 2 or 3
11 a 　i $2 \leqslant x \leqslant 5$　　ii $-2 \leqslant x < 1$　iii $-2 < x \leqslant 3$
　　 iv $\frac{1}{2} \leqslant x < 3$　　v $\frac{-2}{3} < x \leqslant \frac{8}{3}$　vi $-5 \leqslant x \leqslant 5$
　　 vii $0 \leqslant x < 6$　viii $\frac{3}{4} < x \leqslant \frac{7}{5}$
　 b 　i

　　 ii

　　 iii

　　 iv

　　 v

　　 vi

　　 vii

　　 viii

　 c 　i 2, 3, 4, 5　ii -2, -1, 0　iii -1, 0, 1, 2, 3　iv 1, 2
　　 v 0, 1, 2　vi -5, -4, -3, -2, -1, 0, 1, 2, 3, 4, 5
　　 vii 0, 1, 2, 3, 4, 5　viii 1

12 a $2 \leqslant x \leqslant 8$　　b $3 < x < 11$　　c $-2 \leqslant x < 3$
　 d $1 \leqslant x \leqslant 4$　　e $-1 < x < \frac{2}{5}$ (or 0.4)　f $\frac{-3}{2} \leqslant x \leqslant 3$
13 a $-3 < x < 0$　　b $-1 \leqslant x < 2$　c $-7 \leqslant x \leqslant -\frac{1}{3}$
　 d $6.5 < x \leqslant 8$　e $-5 \leqslant x < 5$　f $-11 \leqslant x < 5$
14 a $-4 \leqslant x < 6$　b $-4 < x \leqslant 2$　c $-2 \leqslant x < 4$
　 d $1 < x < 4$　　e $-2 \leqslant x < -1$　f $-3 < x \leqslant 2$
15 -2

2.5 Sequences

1 a i 18, 23　　ii add 5
　 b i 5, 2　　　ii subtract 3
　 c i 8, 12, 14　ii add 2
　 d i 7, 10, 16　ii add 3
　 e i -7, -11, -15　ii subtract 4
　 f i -16, -4, 8　ii add 6
2 a 40, 80; $\times 2$　　　　　b 64, 256; $\times 4$
　 c 100, 50; $\div 2$ or $\times \frac{1}{2}$　d 3, 30; $\times 10$
　 e -5, -1; $\div 5$　　　　f -24, 48; $\times (-2)$
3 a 6, 10　　b -9, -15　c 5, 8
　 d -1, -4　e 1.2, 2.0　f 0.33, 0.55
4 a **2**, **5**, 7, 12, **19**　b **1**, **3**, 4, 7, **11**　c 0, **6**, 6, **12**, 18
　 d **0.5**, **0.5**, 1, 1.5, **2.5**　e 5, **−4**, 1, −3, **−2**　f **−6**, 8, 2, **10**, **12**
5 a 55
　 b No, because in the sequence 21 and 34 are consecutive
　　 terms and the terms are increasing, so it won't go back
　　 to 27.
6 a 2, 4, 6, 8, 10　　　　　b 3, 4, 5, 6, 7
　 c 2, 6, 10, 14, 18　　　d 8, 18, 28, 38, 48
　 e 16, 12, 8, 4, 0　　　f -2, -5, -8, -11, -14
　 g 3.5, 4, 4.5, 5, 5.5　h 11, 16, 21, 26, 31
7 a 6, 46　　b 30, 150　　c 12, 92　　d 51, 251
　 e 60, -100　f -9, -89　g 2, 22　　h 6, 10
8 a $7n - 5$　b $3n + 5$　　c $6n - 1$　d $10 - n$
　 e $-2n + 3$　f $-3n - 4$
9 $2n - 1$
10 a nth term $3n - 4$, yes　　b nth term $6n + 2$, no
　 c nth term $5 - 3n$, no　　d nth term $7n + 2$, yes
　 e nth term $22 - 4n$, no　　f nth term $4n + 3$, yes
11 a 51　　b 51　　c 53　　d 55
12 a $\times 3$; 27, 81　b $\times 5$; 50, 1250
　 c $\times 2$; 12, 48　d $\div 10$; 45, 4.5
13 a June　　b £63
14 a 2, 8, 18, 32, 50　　　b 5, 20, 45, 80, 125
　 c 2, 5, 10, 17, 26　　　d 2, 11, 26, 47, 74
　 e $\frac{1}{2}$, 2, $4\frac{1}{2}$, 8, $12\frac{1}{2}$　　f 99, 96, 91, 84, 75
15 a i 3, 5, 7, 9
　　 ii add the odd numbers starting with 3
　 b i -7, -5, -3, -1
　　 ii subtract the odd numbers starting with 7
16 a Arithmetic (C)　　b Quadratic (D)
　 d Geometric (B)　　d Fibonacci-type (A)

2.6 Straight–line graphs

1 Line A $x = 3$
　 Line B $x = -2$
　 Line C $x = 0$
2 a Line 1 = On Your Bike
　　 Line 2 = Bikes 4 U
　 b On Your Bike (4 hours costs £35, compared with
　　 Bikes 4 U £40)
3 A 3, B 1, C -1, D -4, E $\frac{1}{2}$, F $-\frac{1}{4}$
4 a $y = 2x - 4$, $y = 2x + 3$　b $y = -x + 2$, $y = -2x + 4$
　 c $y = x + 3$, $y = 2x + 3$　d $y = 5x + 1$
　 e $y = \frac{1}{4}$

5 Line A $y = 2x - 1$
   Line B $y = -\frac{1}{2}x + 2$
   Line C $y = \frac{3}{2}x$
   Line D $y = \frac{5}{2}x + 3$
   Line E $y = -3x - 4$

6 a

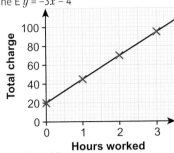

   b $y = 25x + 20$
   c Gradient = hourly rate, y-intercept = callout fee

7 a $y = 5x + 2$
   b $y = 7x - 2$
   c $y = 0.5x + 2.5$
   d $y = -2x + 10$

8 a–d

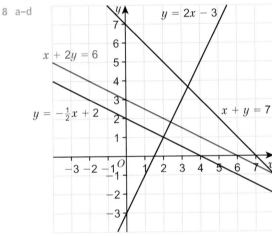

9 a $y = 2x - 2$
   b $y = 3x + 1$
   c $y = \frac{1}{2}x - 3$
   d $y = -2x + 5$
   e $y = \frac{1}{4}x + 5$
   f $y = 2x - 1.5$

2.7 Non-linear graphs

1 a $x = 1$
   b $y = -5$
   c $(1, -6)$

2 a

x	-3	-2	-1	0	1	2	3
y	12	7	4	3	4	7	12

   b

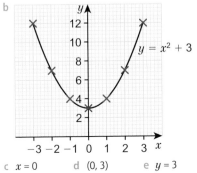

   c $x = 0$ d $(0, 3)$ e $y = 3$

3 a, b

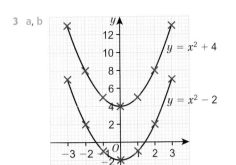

4 a

x	-2	-1	0	1	2	3	4
y	11	5	1	-1	-1	1	5

   b

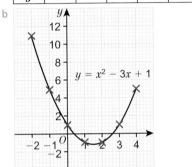

   c i $x = 1.5$ ii $(1.5, -1.25)$ d $x \approx 0.4$ or $x \approx 2.6$

5 a, b

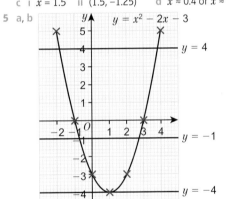

   i $x = -1$ or $x = 3$ ii $x \approx -1.8$ or $x \approx 3.8$
   iii $x \approx -0.7$ or $x \approx 2.7$ iv $x = 1$ (repeated root)

6 a

x	-3	-2	-1	0	1	2	3
y	-30	-11	-4	-3	-2	5	24

   b

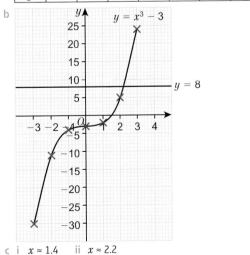

   c i $x \approx 1.4$ ii $x \approx 2.2$

89

7

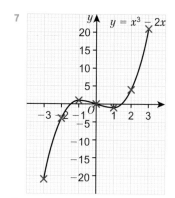
$y = x^3 - 2x$

8 a

x	-4	-3	-2	-1	$-\frac{1}{2}$	$-\frac{1}{4}$	$\frac{1}{4}$	$\frac{1}{2}$	1	2	3	4
$-\frac{1}{x}$	$\frac{1}{4}$	$\frac{1}{3}$	$\frac{1}{2}$	1	2	4	-4	-2	-1	$-\frac{1}{2}$	$-\frac{1}{3}$	$-\frac{1}{4}$

b

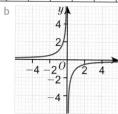

c Students' own answers, e.g. Similarities are that both graphs have the same basic shape, with two parts and the x- and y-axes as asymptotes; both are symmetrical in $y = x$ (and $y = -x$) and their two parts are rotations of each other by 180° about the origin. Differences are that where one graph has positive y-values the other has negative, and vice versa; the two graphs are the reflections of each other in the x-axis (or y-axis).

9 a

x	-4	-3	-2	-1	$-\frac{1}{2}$	$-\frac{1}{4}$	$\frac{1}{4}$	$\frac{1}{2}$	1	2	3	4
$\frac{2}{x}$	$-\frac{1}{2}$	$-\frac{2}{3}$	-1	-2	-4	-8	8	4	2	1	$\frac{2}{3}$	$\frac{1}{2}$

b

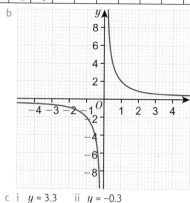

c i $y \approx 3.3$ ii $y \approx -0.3$

2.8 Solving equations

1

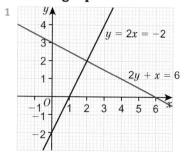
$y = 2x = -2$
$2y + x = 6$

2 a $(x - 10)(x + 10)$ **b** $(x + 3)(x + 1)$
  c $(x - 7)(x + 5)$ **d** $(x - 4)(x - 3)$
3 a $x = \pm7$ **b** $x = \pm9$ **c** $x = \pm3$ **d** $x = \pm11$
  e $x = \pm8$ **f** $x = \pm12$
4 a $x = \pm5$ **b** $x = \pm4$ **c** $x = \pm13$ **d** $x = \pm2$
  e $x = \pm6$ **f** $x = \pm0.5$
5 a $x = -5, x = -3$ **b** $x = 2, x = -6$ **c** $x = -7, x = 3$
  d $x = -9$ **e** $x = 4, x = -8$ **f** $x = 6, x = 3$
  g $x = 2, x = 3$ **h** $x = 4$ **i** $x = -5, x = 8$
6 $x^2 - 10x + 25 = 0$ has solution $x = 5$
  $x^2 + 10x + 25 = 0$ has solution $x = -5$
7 a $x = 6, x = -3$ **b** $x = -1, x = -7$ **c** $x = 8, x = 1$
  d $x = -5, x = -7$ **e** $x = 4, x = 6$ **f** $x = 4, x = -3$
8 $x = 8$; length = 8 cm, width = 2 cm
9 a $x = \pm10$ **b** $x = \pm4$ **c** $x = \pm3$ **d** $x = \pm4$
  e $x = \pm2$ **f** $x = \pm5$
10 $x = 2, y = 2$
11 a $x = 1, y = 3$

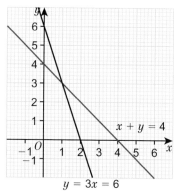
$x + y = 4$
$y = 3x = 6$

  b $x = 0, y = 1$

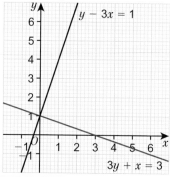
$y - 3x = 1$
$3y + x = 3$

  c $x = -1, y = 2$

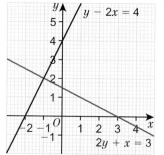
$y - 2x = 4$
$2y + x = 3$

12 a $x = 1, y = 3$ **b** $x = 6, y = 2$ **c** $x = -3, y = 2$
  d $x = 2, y = -1$ **e** $x = -2, y = -5$ **f** $x = 4, y = 10$
13 a $x = 1.2, y = 0.6$ **b** $x = 1.5, y = -2.1$
14 a £6.95   **b** £2.20

2.9 Proof

1 $n, n + 1, n + 7, n + 8$
2 a $x^2 - x - 6$ **b** $x^2 - 5x - 14$ **c** $x^2 - 8x + 15$
3 a equation   **b** identity   **c** identity   **d** equation
4 Expand or factorise to prove the identities.

5   $x(x + 2) + 3(x - 3) = x^2 + 2x + 3x - 9 = x^2 + 5x - 9$

6   $\frac{1}{2} \times x(2x + 6) = x(x + 3) = x^2 + 3x$

7   Volume of triangular prism = $5x^2 + 15x$
    Volume of cuboid = $20x^2 + 60x = 4(5x^2 + 15x)$

8   a $27 - 17 = 10$
    b Students' own answers
    c $(n + 5) + (n + 6) - (n + n + 1) = 2n + 11 - 2n - 1 = 10$

9   $n(n + 7) - (n + 1)(n + 6) = n^2 + 7n - (n^2 + 7n + 6) = 6$

10 a Students' own predictions (5 is correct)
    b $n(n + 6) - (n + 1)(n + 5) = n^2 + 6n - (n^2 + 6n - 5) = 5$

2.10 Mixed exercise

1   a $x^2 = 17$
    b $x = \sqrt{17}$

2   a $n^2 + 14 = 95$
    b $n = 9$

3   a $8bc$
    b $6w - 15t$
    c $x^2 + 5x - 14$

4   $2x(x + 4) - x(2x - 1) = 2x^2 + 8x - 2x^2 + x = 9x$

5   a $6n - 3$
    b No; $6n - 3 = 150$ gives $n = 153 \div 6$, which is not an integer.

6   $x = 3$ cm

7   a
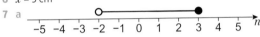
    b $x \geqslant 3.5$

8   $x = 3, y = 5$

9   She has squared each term in the bracket, instead of expanding $(x + 5)(x + 5)$.

10 $x = 6, y = 2$

11 a $A = \frac{1}{2}(a + b)h$
    b $h = \dfrac{2A}{a + b}$
    c $b = \dfrac{2A}{h} - a$

12 $y = 3x - 5$

13 Yes, because the solution is $1 \leqslant x \leqslant 7.5$, and 7 is within this range.

14 a m^{-10} or $\dfrac{1}{m^{10}}$
    b $(x - 2)(x + 5)$

15 $x < 6$ and $x > 4$, so $x = 5$

16 a

x	0.5	1	2	4	5	8
y	8	4	2	1	0.8	0.5

    b
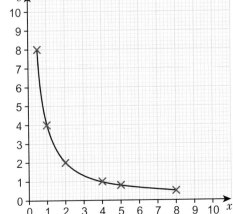

3 RATIO, PROPORTION AND RATES OF CHANGE

3.1 Calculations

1   a £10, £40   b 1.1 litres, 5.5 litres
    c 437.5 g, 262.5 g   d £28.80, £67.20

2   £120

3   £49

4   Bill £6406.25, Trevor £3843.75

5   a 250 g   b 150 g

6   a 2 : 2 : 1   b 1 : 2

7   6, 30, 42

8   a Sunset orange   b Tropical orange

9   a 15 : 1   b 11 : 1   c 60 : 1   d 4.59 : 1
    e 12 : 1   f 15.625 : 1   g 2.25 : 1   h 47 : 1

10 a 82.8 kg of sand, 27.6 kg of cement
    b $1\frac{1}{2}$ kg of cement, $7\frac{1}{2}$ kg of gravel

11 a $\frac{2}{3}$ b $\frac{3}{2}$

12 Squash A 1 : 6, squash B 2 : 15

13 a

x	1	2	3	4
y	3	6	9	12

    b 3 : 1; the ratios are all the same
    c 1 : 3
    d i 5 : 1   ii 1 : 5
    e 1 : 2

3.2 Growth and decay

1   a £54   b 994.5 km

2   a £63.75   b 4.85 kg

3   a £10.50   b 87.5%

4   69%

5   13.1%

6   60%

7   a £275   b £110   c 50p   d £8.50

8   £21

9   £402.94

10 £195 776

11 a £88.26   b £5222.42   c £2532.51
    d £251 753.06

3.3 Compound measures

1   10.5 g/cm³

2   0.8 N/m²

3   19.425 g/cm³

4   7.87 g/cm³

5   4.5 g/cm³

6   565 g

7   84 cm³

8   4 N/cm²

9   4750 N

10 28 cm²

11 The woman exerts the greater pressure.
    (Woman 3.61 N/cm², man 2.66 N/cm²)

3.4 Rates of change

1   a 7 m
    b The water was 7 m deep at the start.
    c 3 m/h
    d The water depth increases by 3 metres per hour.
    e $d = 3t + 7$

2   a 8.45 am   b About 11.42 am   c About 260 km
    d Between 12.30 and 1.30 pm; steepest graph section
    e Train A 87.3… km/h, train B: 91.4… km/h
    f Train B

3 a

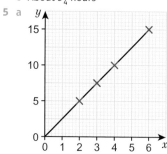

b 72 km/h, 127 km/h, 144 km/h   c 98 km/h

4 a 98 litres
b −30 l/h; 30 litres of water are leaking out per hour.
c About $3\frac{1}{4}$ hours

5 a

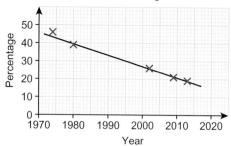

b $y = 2.5x$ c Yes, straight line through (0, 0).

6 a $v = 2t + 3$ b No, not through origin.

7 a The temperature at 8 am is −20 °C.
b 2 pm, when it started getting colder.
c About $5\frac{1}{4}$ hours   d About −5 °C   e About −17 °C

8 a 0 km/h   b 0.2 hours (or 12 minutes)   c 300 km/h²
d i 10 km/h, 67 km/h²   ii 15 km/h, 320 km/h²

9

Percentage of adults in England who smoked

a About 2011   b About 2027
c Answer to part **a** is more reliable as it is within range of the given data set. For part **b**, we don't know that the decrease would continue at the same rate.

3.5 Direct and inverse proportion

1 a $k = 8$ b $k = 16$ c $k = 2.7$ d $k = 4.8$

2 a $F = ka$ b $v = \frac{k}{t}$ c $x = \frac{k}{y}$ d $c = kd$

3 a $k = 4$ b $y = 4x$ c $y = 4$ d $x = 2.5$

4 a $A = 0.4m$ b $m = 12.5$

5 $s = 22.4$

6 a 6 days   b $1\frac{1}{2}$ days   c $1\frac{1}{2}$ days

7 8 hours

8 a $r = \frac{10}{t}$ b $r = 0.5$

9 $v = 0.111\ldots$

10 a

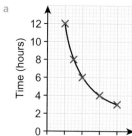

b Inverse proportion
c About 4.8 hours

3.6 Similarity

1 a 6 cm   b 13 cm

2 a i DF   ii AC   iii EF
b i $\frac{3}{2}$ ii $\frac{2}{3}$
c 6 cm   d 8 cm

3 a 12.6 cm
b i 20°   ii 50°

4 a Angle ADB = 60° and angle PQR = 120°, so the shapes have 4 equal angles. AD is 3 × PQ, and AB is 3 × QR, so corresponding sides are in the same ratio.
b 10.5 cm
c 5 cm

5 $x = 12.8$ cm, $y = 7$ cm

6 a 50°   b 12.5 cm   c 2.5 cm

7 a GJ and HI are parallel, so angle FGJ = angle FHI and angle FJG = angle FIH. Both triangles have the same angles.
b 4 cm

8 a 1 cm   b 1.25 cm

9 30 cm

10 9.8 mm

3.7 Mixed exercise

1 a $\frac{1}{4}$ b 11.66… litres, so 12 bottles

2 Jade's drink

3 108 students

4 a 40
b The car is travelling 40 miles on each gallon of petrol.

5 5.2%

6 a

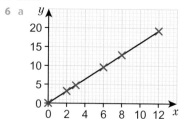

b $y = 1.6x$

7 $r = 2.33$

8 £24

9 1.3×10^7 s

10 1.5 cm

11 £5153.96 − £4800 = £353.96

12 Scheme A: interest = 2500 × 0.04 × 2 = £200.
Scheme B: amount = 2500 × 1.039² = £2698.80; interest = £2698.80 − £2500 = £198.80.
Scheme A gives £1.20 more.

13 £2700

14 7 g/cm³

15 15 625 N/m²

4 PROBABILITY

4.1 Venn diagrams

1 a B′ (C)   b A ∪ B (B)   c A ∩ B (A)   d A ∩ B′ (D)

2 a A = {1, 2, 3, 4, 5}, B = {2, 4, 6, 8, 10},
      ξ = {1, 2, 3, 4, 5, 6, 7, 8, 9, 10, 11}

   b
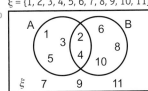

   c i   false
      ii true
      iii true
      iv true
      v   false
      vi false

3 a
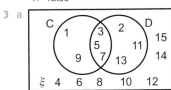

   b i   {3, 5, 7}
      ii   {1, 2, 3, 5, 7, 9, 11, 13}
      iii {2, 4, 6, 8, 10, 11, 12, 13, 14, 15}
      iv   {1, 4, 6, 8, 9, 10, 12, 14, 15}
      v   {1, 9}
      vi   {2, 11, 13}

4 a
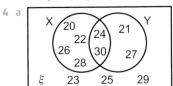

   b i   {21, 23, 25, 27, 29}
      ii   {24, 30}
      iii {20, 22, 23, 25, 26, 28, 29, 30}
      iv   {20, 21, 22, 24, 26, 27, 28 30}
      v   {21, 27}
      vi   {20, 22, 26, 28}

5 a 6 people   b 5 people   c 2 people   d 17 people
   e $\frac{9}{17}$ f $\frac{4}{17}$

6 a
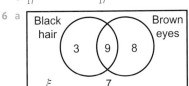

   b $\frac{9}{27}\left(=\frac{1}{3}\right)$ c $\frac{3}{27}\left(=\frac{1}{9}\right)$ d $\frac{7}{27}$

7 a $\frac{3}{35}$ b $\frac{15}{35}\left(=\frac{3}{7}\right)$ c $\frac{2}{35}$

8 $\frac{12}{117}$

   (Venn diagram: Starter, Main, Dessert — 2, 23, 12, 41, 8, 26, 5, ξ)

9 a $\frac{6}{10}\left(=\frac{3}{5}\right)$ b $\frac{5}{10}\left(=\frac{1}{2}\right)$ c $\frac{3}{10}$ d $\frac{8}{10}\left(=\frac{4}{5}\right)$
   e $\frac{4}{10}\left(=\frac{2}{5}\right)$ f $\frac{5}{10}\left(=\frac{1}{2}\right)$ g $\frac{2}{10}\left(=\frac{1}{5}\right)$ h $\frac{3}{10}$

10 a
   (Venn diagram: S, M — 1, 18, 2, ξ 3)

    b i $\frac{19}{24}$ ii $\frac{18}{24}\left(=\frac{3}{4}\right)$ iii $\frac{21}{24}\left(=\frac{7}{8}\right)$ iv $\frac{2}{24}\left(=\frac{1}{12}\right)$

4.2 Frequency trees and tree diagrams

1 a $\frac{1}{6}$ b $\frac{5}{6}$ c $\frac{3}{6}\left(=\frac{1}{2}\right)$ d $\frac{3}{6}\left(=\frac{1}{2}\right)$
   e $\frac{2}{6}\left(=\frac{1}{3}\right)$ f $\frac{4}{6}\left(=\frac{2}{3}\right)$

2
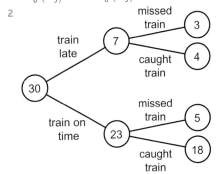

3
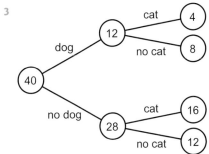

4 a
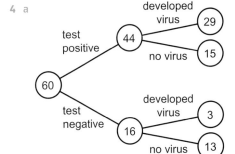

   b 3 people

5 a
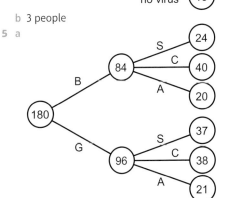

   b 78 students
6 7 people
7 a $\frac{9}{49}$ b $\frac{16}{49}$ c $\frac{24}{49}$ d $\frac{25}{49}$

8 a

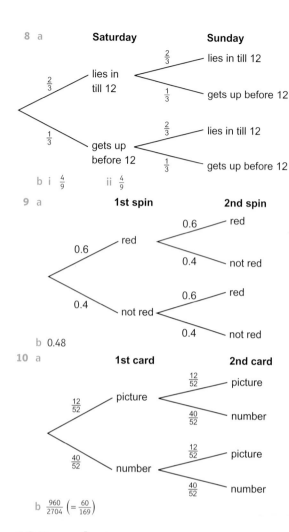

Saturday   **Sunday**

$\frac{2}{3}$ lies in till 12 — $\frac{2}{3}$ lies in till 12

$\frac{1}{3}$ gets up before 12

$\frac{1}{3}$ gets up before 12 — $\frac{2}{3}$ lies in till 12

$\frac{1}{3}$ gets up before 12

  b i $\frac{4}{9}$ ii $\frac{4}{9}$

9 a

1st spin   **2nd spin**

0.6 red — 0.6 red

0.4 not red

0.4 not red — 0.6 red

0.4 not red

  b 0.48

10 a

1st card   **2nd card**

$\frac{12}{52}$ picture — $\frac{12}{52}$ picture

$\frac{40}{52}$ number

$\frac{40}{52}$ number — $\frac{12}{52}$ picture

$\frac{40}{52}$ number

  b $\frac{960}{2704}\left(=\frac{60}{169}\right)$

4.3 Dependent events

1 a $\frac{4}{12}\left(=\frac{1}{3}\right)$ b $\frac{8}{11}$

2

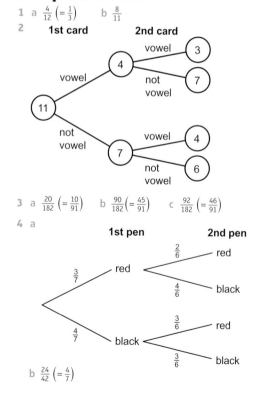

1st card   **2nd card**

11 — vowel — 4 — vowel — 3

               not vowel — 7

   not vowel — 7 — vowel — 4

               not vowel — 6

3 a $\frac{20}{182}\left(=\frac{10}{91}\right)$ b $\frac{90}{182}\left(=\frac{45}{91}\right)$ c $\frac{92}{182}\left(=\frac{46}{91}\right)$

4 a

1st pen   **2nd pen**

$\frac{3}{7}$ red — $\frac{2}{6}$ red

$\frac{4}{6}$ black

$\frac{4}{7}$ black — $\frac{3}{6}$ red

$\frac{3}{6}$ black

  b $\frac{24}{42}\left(=\frac{4}{7}\right)$

5 a

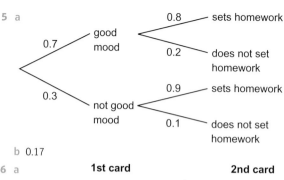

0.7 good mood — 0.8 sets homework

              0.2 does not set homework

0.3 not good mood — 0.9 sets homework

              0.1 does not set homework

  b 0.17

6 a

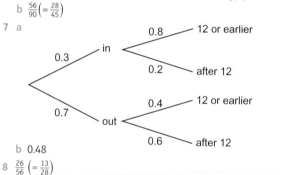

1st card   **2nd card**

$\frac{2}{10}$ multiple of 4 — $\frac{1}{9}$ multiple of 4

              $\frac{8}{9}$ not multiple of 4

$\frac{8}{10}$ not multiple of 4 — $\frac{2}{9}$ multiple of 4

              $\frac{7}{9}$ not multiple of 4

  b $\frac{56}{90}\left(=\frac{28}{45}\right)$

7 a

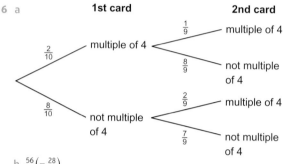

0.3 in — 0.8 12 or earlier

        0.2 after 12

0.7 out — 0.4 12 or earlier

        0.6 after 12

  b 0.48

8 $\frac{26}{56}\left(=\frac{13}{28}\right)$

9 a True (because train late, P(late to work) = 0.6, train on time, P(late to work) = 0.1, and 0.6 > 0.1)
  b False (because P(early) = 0.4, P(not early) = 0.6, and 0.4 < 0.6)
  c False (because P(late) = 0.12 + 0.08 = 0.2, and 0.2 < 0.25 = $\frac{1}{4}$)
  d True (because P(early) = 0.02 + 0.32 = 0.34; 10 × 0.34 = 3.4, and 3.4 < 4)

10 a

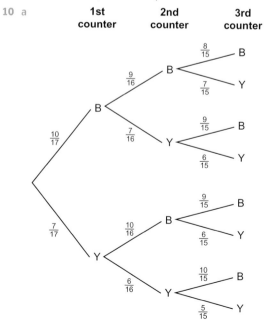

1st counter   **2nd counter**   **3rd counter**

$\frac{10}{17}$ B — $\frac{9}{16}$ B — $\frac{8}{15}$ B

                  $\frac{7}{15}$ Y

        $\frac{7}{16}$ Y — $\frac{9}{15}$ B

                  $\frac{6}{15}$ Y

$\frac{7}{17}$ Y — $\frac{10}{16}$ B — $\frac{9}{15}$ B

                  $\frac{6}{15}$ Y

        $\frac{6}{16}$ Y — $\frac{10}{15}$ B

                  $\frac{5}{15}$ Y

b i $\frac{7}{136}$ ii $\frac{63}{136}$

c 2 blues and a yellow $\left(\frac{63}{136} > \frac{42}{136}\right)$

4.4 Mixed exercise

1 a A = {1, 2, 4, 8, 16},
    B = {1, 2, 3, 6, 9, 18},
    C = {2, 4, 6, 8, 10}

  b i {1, 2}   ii {2, 6}   iii {2, 4, 8}   iv {2}

  c

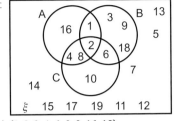

  d {1, 2, 3, 4, 6, 8, 9, 16, 18}

2 a

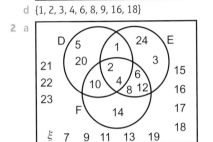

  b {1, 2, 4, 5, 6, 8, 10, 12, 14, 20}

  c 16

3 a

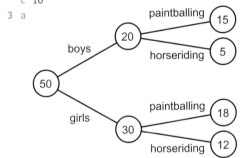

  b 33 children

4 11 children

5 29 girls

6 51 children

7 53 male students

8 a

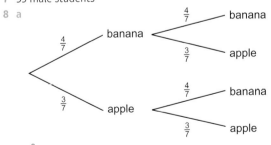

  b $\frac{9}{49}$

9 a

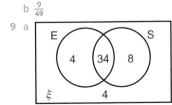

  b No, because 4 students failed, and $\frac{4}{50}$ is 8% not 4%.

10 a

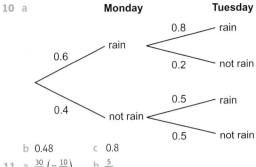

  b 0.48   c 0.8

11 a $\frac{30}{63}\left(=\frac{10}{21}\right)$ b $\frac{5}{63}$

12 a

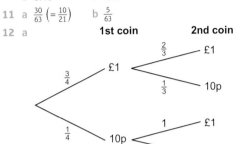

  b $\frac{6}{12}\left(=\frac{1}{2}\right)$ c $\frac{1}{2}$

13 a

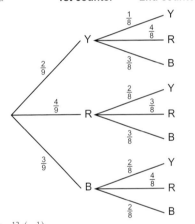

  b $\frac{12}{72}\left(=\frac{1}{6}\right)$

14 $\frac{8}{15}$

15 $\frac{22}{30}\left(=\frac{11}{15}\right)$

5 GEOMETRY AND MEASURES

5.1 Problem-solving

1 a $a = 110°$, $b = 110°$, $c = 70°$

  b $d = 125°$

  c $e = 40°$, $f = 50°$

2 a 40°   b 140°

3 a 45°   b 67.5°   c 67.5°

4 a

  b All angles are 60°.

  c Equilateral because all angles are 60°.

5 $x = 50°$, $2x = 100°$, $x - 30° = 20°$, $2x + 10° = 110°$, $y = 80°$

6 a She has missed out the 90° angle in the calculation, and found x, not the angles.

  b $8x + 60° = 900°$, so $x = 105°$.
    Angles are 90°, 105°, 120°, 150°, 145°, 140°, 150°

7 a $a = b = 120°$ (interior angles in regular hexagon),
   $c = 30°$ (base angle in isosceles triangle)
  b $\angle C = \angle D$ and BC = DE (interior angles and sides of
   regular hexagon), so BCDE is an isosceles trapezium.
  c 60°

8 a $a = 135°$ (interior angle of regular octagon),
   $b = 135° - 90° = 45°$,
   $c = 78°$ (angles in a triangle sum to 180°),
   $d = 78°$ (corresponding and vertically opposite angles),
   $e = 78°$ (same steps as for a to d),
   $f = 24°$ (angles in a triangle sum to 180°)

9 a Exterior angle = $30° \Rightarrow 12$ sides, $y = 360 \div 12 = 30°$.
   Angle y is the same size as the exterior angle.
  b $z = 20°$. Angle $20° \Rightarrow 18$ sides and $360° \div 18 = 20°$

10 $n = 10$

11 10 sides ($x = 36°$)

12 Regular octagon

5.2 Transformations

1

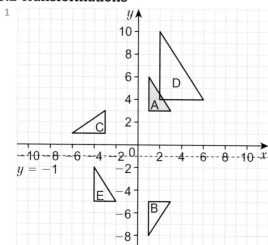

2 a–d

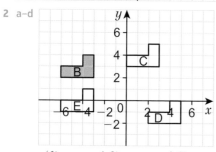

  e i $\binom{6}{4}$ ii $\binom{8}{-4}$ iii $\binom{0}{-3}$
   iv $\binom{-2}{5}$ v $\binom{8}{-1}$ vi $\binom{0}{3}$

3 a–c

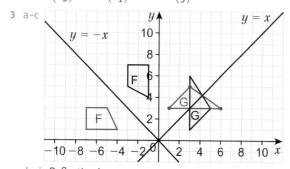

  d i Reflection in $y = x$
   ii Reflection in $y = -x$
  e Rotation of 90° anticlockwise about (−2.5, −2.5)

4 a–c

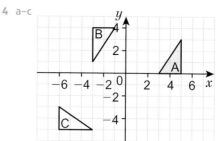

  d 90° clockwise about (−3, 3)

5 a–c

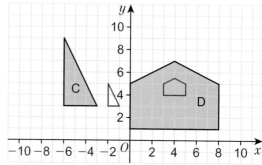

  d i Enlargement by a scale factor of $\frac{1}{4}$ from centre (3, −5)
   ii Enlargement by a scale factor of 4 from centre (3, −5)

6 a Rotation of 180° about (−1, −3)
  b

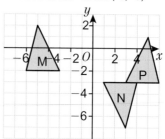

  c Rotation of 180° about (3.5, −3)

7 a Translation $\binom{0}{-6}$; reflection in $y = 1$
  b Reflection in $y = x$; rotation of 90° anticlockwise about
   (4, 4)
  c

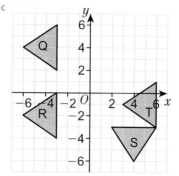

  d i Translation $\binom{-9}{-1}$
   ii $\binom{9}{-5}$

8 a, b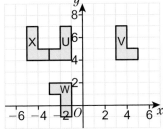

c Rotation of 180° about (1, 3)
d $k = -3$

9 a–c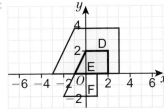

d Enlargement by a scale factor of $\frac{1}{2}$ from centre (−1, −4)
10 Students' own answers (any enlargement with scale factor ≠ 1 gives a non-congruent image)

11 a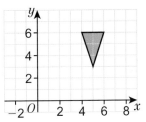

b Students' own answers

5.3 Vectors

1 a $\begin{pmatrix} 2 \\ 3 \end{pmatrix}$ b $\begin{pmatrix} 0 \\ 4 \end{pmatrix}$ c $\begin{pmatrix} -4 \\ 2 \end{pmatrix}$ d $\begin{pmatrix} -1 \\ -5 \end{pmatrix}$

2 a $\begin{pmatrix} 5 \\ 5 \end{pmatrix}$ b $\begin{pmatrix} -1 \\ 4 \end{pmatrix}$ c $\begin{pmatrix} 3 \\ 5 \end{pmatrix}$ d $\begin{pmatrix} -5 \\ 2 \end{pmatrix}$

3 a i $\begin{pmatrix} 4 \\ 1 \end{pmatrix}$ ii $\begin{pmatrix} 2 \\ 5 \end{pmatrix}$ iii $\begin{pmatrix} 6 \\ 6 \end{pmatrix}$

b $\begin{pmatrix} 4 \\ 1 \end{pmatrix} + \begin{pmatrix} 2 \\ 5 \end{pmatrix} = \begin{pmatrix} 6 \\ 6 \end{pmatrix}$

4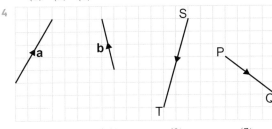

5 a $\begin{pmatrix} 3 \\ 2 \end{pmatrix}$ b $\begin{pmatrix} 3 \\ -1 \end{pmatrix}$ c $\begin{pmatrix} 2 \\ 1 \end{pmatrix}$ d $\begin{pmatrix} 7 \\ 2 \end{pmatrix}$

e $\begin{pmatrix} -3 \\ -1 \end{pmatrix}$ f $\begin{pmatrix} 4 \\ 6 \end{pmatrix}$

6 $\begin{pmatrix} 1 \\ 5 \end{pmatrix}$

7 a, b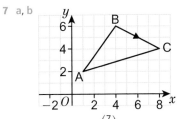

c (8, 4) d $\begin{pmatrix} 7 \\ 2 \end{pmatrix}$

8 a (2, 7) b $\begin{pmatrix} 2 \\ 6 \end{pmatrix}$

9 a $\begin{pmatrix} -3 \\ 6 \end{pmatrix}$ b $\begin{pmatrix} -8 \\ 0 \end{pmatrix}$ c $\begin{pmatrix} 7 \\ 0 \end{pmatrix}$ d $\begin{pmatrix} 1 \\ -4 \end{pmatrix}$

10 a $\begin{pmatrix} -5 \\ -3 \end{pmatrix}$ b $\begin{pmatrix} -10 \\ -6 \end{pmatrix}$ c $\begin{pmatrix} 2 \\ -5 \end{pmatrix}$ d $\begin{pmatrix} -7 \\ 2 \end{pmatrix}$

e $\begin{pmatrix} 5 \\ -6 \end{pmatrix}$ f $\begin{pmatrix} -6 \\ 15 \end{pmatrix}$ g $\begin{pmatrix} -6 \\ 2 \end{pmatrix}$ h $\begin{pmatrix} -11 \\ -1 \end{pmatrix}$

11 a **a** + **b** b **b** + **c** c **a** + **b** + **c** d −**b** − **a**

12 a i −**u** ii **u** + **v** iii −**v** iv −**u** + **w**

b $\overrightarrow{QR}$ = −**v** − **u** + **w** = **w** − **u** − **v**

−$\overrightarrow{RQ}$ = −(−**w** + **u** + **v**) = **w** − **u** − **v** = $\overrightarrow{QR}$

5.4 Right-angled triangles

1 a $c = 5.8$ b $b = 7.7$ c $x = 4.6$ d $x = 5.3$
e $\theta = 59.0°$ f $\theta = 16.6°$

2 a 8.60 cm b 5.24 cm

3 a $x = \sqrt{39}$ cm b $y = 2\sqrt{30}$ cm c $z = 2\sqrt{21}$ cm

4 EF = $\sqrt{65}$ (AB = $\sqrt{32}$, CD = $\sqrt{40}$)

5 a $x = 6.9$ b $y = 22.0$ c $z = 13.6$

6 a 35° b 58° c 34° d 53°

7 a $d = 38.5$ mm b $e = 4.3$ m c $f = 28.4$ km

8 15.39 m

9 71.5 m

10 a

b i $\frac{1}{2}$ ii $\frac{1}{\sqrt{2}}$ iii $\frac{\sqrt{3}}{2}$ iv $\frac{\sqrt{3}}{2}$ v $\frac{1}{\sqrt{2}}$
vi $\frac{1}{2}$ vii $\frac{1}{\sqrt{3}}$ viii 1 ix $\sqrt{3}$

c i sin 0° = cos 90° = tan 0° = 0
ii cos 0° = sin 90° = 1

11 a cos 60° = $\frac{1}{2}$ = sin 30° b cos 45° = $\frac{1}{\sqrt{2}}$ = sin 45°

5.5 Constructions, loci and bearings

1 a Perpendicular bisector of 10 cm line accurately constructed
b Accurate 70° angle drawn and bisected, showing construction arcs

2 a Perpendicular from point P to line accurately constructed
b Perpendicular at point C on line accurately constructed

3 a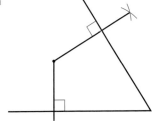

b South; this is the shorter distance.

4 Diagram accurately traced and angle bisector constructed.

5

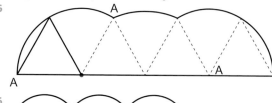

6

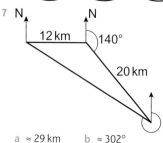

7

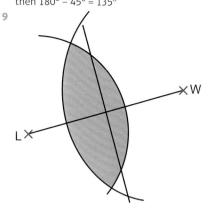

a ≈ 29 km   b ≈ 302°

8 Construct 135° by first constructing 90° and its bisector, then 180° − 45° = 135°

9

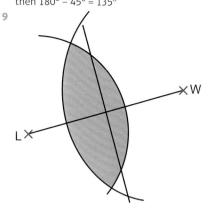

5.6 Perimeter, area and volume

1 a 113.1 cm²   b 37.7 cm
2 a 226 cm³   b 180 cm³
3 a 4.46 cm²   b 2.8 cm
4 33.1 cm²
5 No, because area of flowerbed = 2.54 m², which is less than one third of 12 m²
6 can B (surface area 277.1 cm² < 324.4 cm²)
7 sector B (B = 8.33…π, A = 8.16…π, C = 7.11…π)
8 3732 cm³
9 377 180 mm³
10 41.5 m²

5.7 Congruence

1 a SAS   b RHS   c ASA   d SSS
2 a Yes, SSS   b Yes; RHS or SSS   c Yes; ASA
d Not congruent
3 a FE = AC = 6 cm and AB = DF = 8 cm; RHS or SSS
b ∠G = ∠H = ∠J = ∠L = 70° and JL = GH; ASA
c ∠Q = ∠O = 100°; SAS
4 Yes; ∠Y = ∠T, ∠XVY = ∠TVU, ∠U = ∠X; SAS or ASA
5 No. Although the angles are same, the two 5 cm sides are not corresponding sides. Corresponding sides not equal.

6 a $x = 5.4$ cm, $y = 3.6$ cm   b $a = 6$ cm, $b = 10.4$ cm
7 a, b

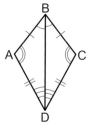

c Triangles ABD and BCD are congruent; SSS or SAS or ASA

5.8 Mixed exercise

1 £269
2

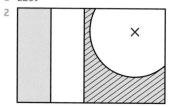

3 39°
4 ∠DEA = 38° (alternate angles)
∠EDA = ∠EAD (base angles of an isosceles triangle)
∠EDA = $\frac{1}{2}$(180 − 38) = 71°
x = 180° − 71° = 109° (angles on a straight line sum to 180°)
5 a, b Accurate construction of original course and
perpendicular bisector.
c 100°, 28 m
6 $a = 49°, b = 63°$
7 Exterior angle of a regular pentagon = $\frac{360}{5}$ = 72°
∠FCD = ∠CDF = 72°
∠CFD = 180° − 72° − 72° = 36°
8 a Translation $\begin{pmatrix} -6 \\ 0 \end{pmatrix}$

b Rotation of 180° about (1, 1)
c Rotation of 180° about (4, 1)
9 Divide the hexagon into triangles to give 6 equilateral
triangles, so interior angles of each triangle are 3 × 60°.
Angle at vertex of hexagon = 2 × 60° = 120°.
Divide the octagon into triangles to give 8 isosceles
triangles, so interior angles of each triangle are 1 × 45°
(at the centre) and 2 × 67.5°.
Angle at vertex of octagon = 2 × 67.5° = 135°.
Angles at a point sum to 360°, so
x = 360° − 120° − 135° = 105°
10 a (9, 1)   b 8 cm   c $\begin{pmatrix} 7 \\ -4 \end{pmatrix}$
11 54°
12 a 25.1 cm   b 2.5 cm per minute
13 a 7.5 km   b 217°
14 a **b** + 2**a**   b **a** − **b**   c **a** + **b**
15 302 cm³
16 a ∠CAB = 180° − 100° = 80° (angles on a straight line sum
to 180°),
∠GCD = 80° (corresponding angles),
∠ACB = 50° (angles on a straight line sum to 180°).
∠ABC = ∠BCD = 50° (alternate angles),
so ∠ACB = ∠ABC, so triangle ABC is isosceles.
b ABCD has two pairs of parallel sides so it is a
parallelogram. Hence AC = BD and AB = CD. Side BC is
common; SSS.
c Triangle ABC is isosceles so AC = AB. Also AC = BD and
AB = CD. So ABCD is a parallelogram with all sides equal
in length, i.e. a rhombus.

6 STATISTICS

6.1 Scatter graphs

1   a   Negative correlation   b   smaller   c   8 km

2   a, c

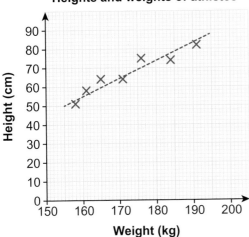

Heights and weights of athletes

   b   Positive correlation
   d   About 73 kg

3   a–c

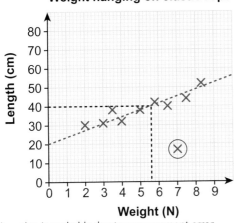

Weight hanging on elastic rope

   b   The value is probably due to measurement error.
   d   About 5.6 N
   e   About 20 cm

4   a–b

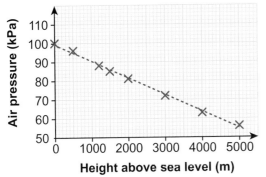

Air pressure and height above sea level

   c   About 82–84 kPa
   d   About 3200–3300 m

5   No correlation

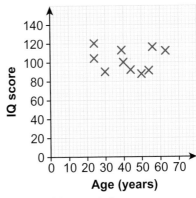

Age and IQ score

6   a   Yes, positive correlation

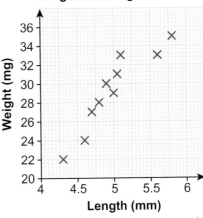

Length and weight of beetles

   b   Beetle A is likely to be the same species, but Beetle B is not (its data point is an outlier).

7   a   Yes, positive correlation
   b   No, because height is unlikely to affect number of films watched.

8   a

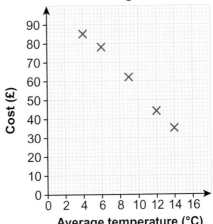

Heating costs

   b   i   About £10   ii   About £55
   c   The estimate for 10°C, because it is interpolation (inside the range of the data given).

9   a   Graph 1, because the points lie closer to line of best fit
   b   Yes (Graph 1), because being good at French could make you more likely to be good at Spanish, or vice versa.

10   a   Negative correlation (C)
   b   No correlation (A)
   c   Positive correlation (B)

6.2 Mean, median, mode and range

1 a

Height, h (m)	Frequency, f	Midpoint of class, m	$m \times f$
$0 < h \leqslant 2.0$	8	1	8
$2.0 < h \leqslant 4.0$	12	3	36
$4.0 < h \leqslant 6.0$	15	5	75
$6.0 < h \leqslant 8.0$	7	7	49
$8.0 < h \leqslant 10.0$	8	9	72
Total	50	**Total**	240

   b 4.8 m

2 9.8 seeds

3 a 4 eggs   b 7 eggs

4 a 150 calls   b $6 < t \leqslant 10$ c 6 minutes   d 73 calls

5 a No, because the mean value should represent the data, but most of the data values are less than her mean value, 15.7 cm.
   b 10.125 cm
   c The mean is an estimate because she used the midpoint of each class as an estimate for all the lengths in the class. The range is an estimate because she didn't know the actual longest and shortest lengths.

6 a 49 cars   b $35 < s \leqslant 40$
   c $35 < s \leqslant 40$ d 34.925 mph
   e No, because all these averages (mean, median, mode) are $\leqslant 40$.

7 a Modal class
   b Mean (61.7) or class containing the median
   c 100; it is unlikely that there were marks of 0 or 100

8 Yes, it went down to 6.2 minutes.

6.3 Sampling

1 a It would be too expensive and time consuming to ask all 80 000 people.
   b 6000 people
   c 6400 people

2 No, because his sample is only in one place, at one time: people who shop mainly online, at out-of-town retail parks or not on Saturdays, won't be included.

3 Most people under 21 don't have a landline so their names won't be in the phone book.

4 Method B, as it is a random sample. Method A is not random as names at the start of the alphabet are more likely to be picked. Method C is not random as slower or harder working students are likely to be picked.

5 Students' own answers, e.g. select seat numbers at random by picking letters from a set of A–Z cards and numbers from a set of 1–20 cards, or select ticket numbers at random by picking stubs out of a hat or using a random number generator.

6 a Biased: only one time, only patients using the service.
   b Biased: only telephone owners, no one from outside these four towns.
   c Biased: people at a rock concert are more likely to choose rock.
   d Unbiased

7 a $\frac{2}{3}$ b $\frac{1}{3}$ c 8 males, 4 females

8 a $\frac{240}{400} = \frac{3}{5}$ b 48 students

9 a 6 employees   b 9 employees

10 a 400 people   b i 150 people   ii 250 people

6.4 Mixed exercise

1 Survey will be biased as only men are asked, and near a bus station so will favour bus passengers. Also, the sample size is too small.

2 Method B will be more representative of all electors. Method A would probably get more young people, including some who are too young to vote.

3 488.75 miles

4 a Point plotted at (6, 35)
   b Negative correlation
   c Line of best fit drawn on graph
   d About 21–26 hot drinks

5 a $0 \leqslant r < 2$ b $0 \leqslant r < 2$
   c

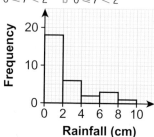

   d Summer, because most days had no or little rainfall

6 a Point plotted at (47, 34)
   b Positive correlation
   c About 48–51 cm

7 a $25\,000 \leqslant s < 30\,000$
   b $30\,000 \leqslant s < 40\,000$
   c £36 236
   d The class containing the median indicates the highest salaries.

8 a $5 \leqslant t < 10$
   b

Time, t (min)	Frequency
$0 \leqslant t < 5$	24
$5 \leqslant t < 10$	50
$10 \leqslant t < 15$	42
$15 \leqslant t < 20$	16

   c 9.4 minutes
   d $5 \leqslant t < 10$
   e 20 minutes

9 6 women

10 The modal class and the class containing median are both $1.5 \leqslant h < 2$, so the claim may or may not be correct. The mean is 1.77 m so claim is correct.

11 a 42 years   b $40 < a \leqslant 50$
   c 2 people   d 5 people